Oregon's Island in the Sky

Oregon's Island in the Sky

Geology Road Guide to Marys Peak

Robert J. Lillie, Ph.D.
Certified Interpretive Trainer
Emeritus Professor of Geosciences
Oregon State University

Wells Creek Publishers
Philomath, Oregon

← Mount Hood from Marys Peak. When clouds fill the Willamette Valley, the "island in the sky" metaphor of Marys Peak comes to life.

Designed by Robert J. Lillie.

Photos and illustrations (unless otherwise noted) by Robert J. Lillie.

ISBN 978-1-540-61196-3

Published by Wells Creek Publishers, LLC, Philomath, Oregon.

Printed by CreateSpace.

Available from Amazon.com and other retail outlets.
www.amazon.com/dp/1540611965

Manufactured in the United States of America.

Published 2017.

Wells Creek Publishers, LLC
22370 Wells Creek Road
Philomath, OR 97370
Phone: 541-231-2247
E-mail: foxglove@peak.org
Web: www.robertjlillie.com/wellscreekpublishers

Cover Photo: Marys Peak from Fitton Green Natural Area near Corvallis, Oregon. The peak has been called "Mouse Mountain," perhaps an even more appropriate name in modern times, as the mountain has the form of a computer mouse.

To my son Ben. You grew up in the shadow of Marys Peak and became a storyteller.

A young Ben Lillie on Marys Peak.

Other books by Robert J. Lillie

Beauty from the Beast: Plate Tectonics and the Landscapes of the Pacific Northwest, Philomath, Oregon: Wells Creek Publishers, 92 pp., 2015. www.amazon.com/dp/1512211893

Parks and Plates: The Geology of Our National Parks, Monuments, and Seashores, New York: W. W. Norton and Company, 298 pp., 2005. www.amazon.com/dp/0134905172

Whole Earth Geophysics: An Introductory Textbook for Geologists and Geophysicists, Englewood Cliffs, N.J., Prentice Hall, Inc., 361 pp., 1999. www.amazon.com/dp/0393924076

Acknowledgments

Marys Peak has been a source of beauty and inspiration to my family, friends and colleagues for most of my life. I am grateful to so many people for sharing outdoor experiences on this amazing mountain, and for countless discussions about its scientific and spiritual significance. *Oregon's Island in the Sky* evolved over many years of visiting a place that is now just minutes from my doorstep.

My colleagues and students at Oregon State University provided much knowledge incorporated into the book. Chris Goldfinger and Bob Lawrence conducted geological mapping of the Corvallis Fault and surrounding region that provided the basis for the simplified geological maps and cross section developed for the book. Bob Yeats and his students advanced understanding of the types of earthquakes in the Pacific Northwest, and on the potential for damaging earthquakes along the Cascadia Subduction Zone and Corvallis Fault. Discussions over the years with Alan Niem and Lockwood DeWitt helped me appreciate details of rock outcroppings and sedimentary processes seen along Highway 34 and the Marys Peak Road.

I am especially grateful to natural and cultural history experts for their inspiring hikes, workshops and public presentations over the years. Jason O'Brien and the Oregon Master Naturalist Program provided the opportunity to work with an amazing group of trainees and engage them during field trips to Marys Peak. Barry Wulff and the Marys Peak Chapter of the Sierra Club provided many thoughtful and inspiring hikes on the peak. Members of the U.S. Forest Service, particularly Lisa Romano, David Thompson and Brian Hoeh, were so helpful in their dedication to stewardship and education on Marys Peak. A special thanks to Ranger Bob Brant for all he does to make Marys Peak such a safe and friendly place to visit.

Many individuals from the Marys Peak Alliance provided resources, encouragement and a wealth of knowledge that helped the book come to fruition. Dave Eckert's tireless leadership and community service efforts are especially appreciated! Stewart Holmes, Susan Van Laere, Daniele McKay, Lee Sherman, Bill Gellatly, Barry Wulff, Bill O'Conner and Dave Eckert provided peer reviews and discussions that greatly improved the original manuscript and illustrations. A special thank you to Phil Hays for his wonderful panoramic views from the summit of Marys Peak and permission to use them.

And I'm so grateful to my wife Barb for our many walks on Marys Peak and her artful eye as the book developed.

Contents

Introduction

Marys Peak
An Island in the Sky

Close your eyes and imagine you're on an island rising high above the clouds of western Oregon. The tree-covered peaks below extend westward to the waters of the Pacific Ocean. To the east lie the Willamette Valley, and then the snow-capped volcanic giants—Mount Hood, Mount Jefferson, and the Three Sisters. Marys Peak is like that island in the sky. A drive to its summit reveals how the Pacific Northwest formed in the distant past, and how Oregon's landscape, ecology and human history continue to develop today.

The beautiful landforms of western Oregon and Washington—and the danger of earthquakes and volcanic eruptions—are products of the Cascadia Subduction Zone. As the Juan de Fuca Plate dives beneath the edge of North America, sedimentary and volcanic layers are scraped off the ocean floor and lifted out of the sea, forming the Coast Range. Farther inland the top of the diving plate becomes so hot that it generates magma that rises and forms the Cascade Volcanoes. The flanks of Marys Peak have rocks that tell the story of these ongoing geological processes, and its summit is one of the premier places in the world to see the full landscape of an active subduction zone.

← Marys Peak, viewed from Fitton Green Natural Area in Corvallis.

1

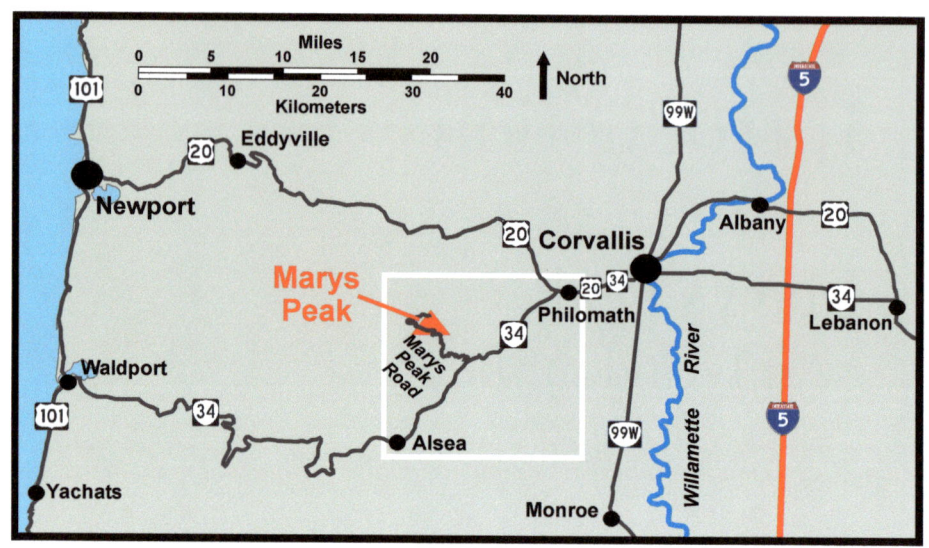

Marys Peak is accessed from the Marys Peak Road, about 15 miles southwest of Corvallis on Oregon Highway 34. The white box encloses the area of the geologic map with road stops on page 8.

Marys Peak Geology Questions

Marys Peak is truly an island in the sky. It towers nearly 500 feet above all the other mountains in the Oregon Coast Range. In its upper reaches it preserves an ecosystem that represents the last vestiges of what western Oregon was like during the last ice age. But how did Marys Peak get to be so high, and why does it remain so much higher than the surrounding terrain? What will it be like in the future, as both natural and human forces modify this fragile environment? And why has this high ground been such a special place to Native Americans and those who came afterward? Answers to these and other questions about Marys Peak involve the plate-tectonic forces that formed the mountain ranges of the Pacific Northwest, and the nature of the geologic layers that are so dramatically seen during a drive, bicycle ride or hike up the mountain.

Oregon's Island in the Sky

The view of Mount Hood across the Willamette Valley emphasizes the metaphor of Marys Peak as an island in the sky (left). When the valley is filled with clouds, Marys Peak and other Coast Range mountains poke up through a sea of white (right), much like the San Juan Islands and Olympic Peninsula rise above the waters of Washington's Puget Sound.

1. What is the significance of an "Island in the Sky?"

The expression "island in the sky" is not unique to Marys Peak. A plateau region with commanding vistas in Utah's Canyonlands National Park is called "Island in the Sky." Isolated mountains of cooler climate, higher moisture and lush vegetation extending high above the dry desert of Arizona, California and Nevada are called "sky islands." Because Marys Peak rises far above the landscape of the Oregon Coast Range and has such unique climate and ecology, it is indeed an island in the sky.

Marys Peak is a very special place to many people, past and present. Native Americans have used its slopes and summit for vision quests for bringing a child into adulthood. Early settlers to the Willamette Valley grazed animals and logged on the peak. From 1946 to 1984, the summit meadows were used for an annual trek sponsored by the Shriners Club that attracted up to 12,000 people per day. During the 20th century a rope tow enabled downhill skiing and the U.S. Forest Service erected an observation tower on the summit (both no longer there). Numerous individuals, families and organizations continue to use the peak for hiking and other outdoor recreation and education. Local middle schools use Marys Peak for a capstone field experience that, like the Kalapuya vision quests, are an important part of the transition of children into adulthood.

The summit area of Marys Peak is a unique ecosystem of noble fir forests and sub-alpine meadows remaining from the last ice age. Climate change and human overuse in recent decades have altered this fragile environment, shrinking the ecological island in the sky.

2. Why are there two parallel mountain ranges in the Pacific Northwest?

Plate tectonics plays a key role. A piece of Earth's outer shell, the Juan de Fuca Plate, lies off the coast of Washington, Oregon and northern California. As the plate dives (subducts) beneath the edge of North America, sedimentary and volcanic rocks are scraped off the ocean floor, forming the Olympics and other coastal mountain ranges. Farther east, the top of the Juan de Fuca Plate reaches depths where it's so hot and the pressure is so great that rocks on top of the plate undergo chemical changes (metamorphism), releasing water in the process. The superheated water causes surrounding rock to melt, forming molten magma. Some of the magma rises to the surface, erupting as lava and forming the Cascade Volcanoes. The region between the two developing mountain ranges remains just above or just below sea level. This is the Willamette Valley of Oregon and the Puget Sound region of Washington, where most of the population of the Pacific Northwest resides. A drive up Marys Peak provides a unique opportunity to view details of the rock layering and landscape of an active zone of plate convergence, the Cascadia Subduction Zone.

Summit of Marys Peak

The top of Marys Peak is special because it offers a 360-degree view of the landscape of the Pacific Northwest. That same perspective is also ideal for the siting of communication towers that stand out on the summit meadows.

3. Why is Marys Peak the highest point in the Oregon Coast Range?

The answer lies in understanding how the mountain formed by examining the rocks exposed on its summit and flanks. The rocks of the Coast Range were manufactured in the ocean, but have been thrust upward and eroded as the ocean floor subducted beneath North America. Typical elevations in the Oregon Coast Range vary from a few hundred feet at passes to 1,500 to 3,500 feet on mountaintops. Marys Peak extends to an unusually high elevation (4,097 feet) for two reasons: first, movement along the Corvallis Fault uplifted oceanic crustal rock (basalt) and its sedimentary cover (sandstone and shale) and, second, a thick, intrusive layer of very hard rock (gabbro) has resisted erosion, so that the summit area of Marys Peak remains high.

4. How does the landscape of western Oregon and Marys Peak affect biology, ecology and human history?

Air from storms arriving from the Pacific Ocean rises and expands over the Coast Range and Cascade mountains, dropping much of its moisture as rain and snow. The dried-out air compresses as it moves down the east slope of the Cascades, creating the high-desert climate of central and eastern Oregon. Oregon thus has a lush "green" western side, and a drier, "brown" side on the east. This varied landscape greatly shaped human history and settlement patterns. Native American populations thrived on abundant hunting and gathering opportunities in the Willamette Valley. When European settlers migrated along the Oregon Trail in the mid-1800s, they generally bypassed the drier regions of eastern Oregon for the lush farmland, easy transportation and relatively mild climate offered by the Willamette Valley.

Dramatic climate effects are also seen on drives up high mountains

The nearly vertical wall of gabbro at Parker Creek Falls provides a wet, shaded environment that promotes the growth of ferns, mosses and lichens. It reveals how chemical and biological activity can eventually wear away even extremely strong rock like the gabbro capping Marys Peak.

such as Marys Peak. These changes are due to lower temperatures and higher precipitation with increasing elevation, as well as changes in the type of bedrock and resulting soils. Marys Peak is a special type of Coast Range mountain known as a "bald." The sub-alpine meadows and spectacular early summer wildflower displays are due to the high elevation and the weathering properties of the gabbro rock capping the mountain.

5. When and where will the next big earthquake occur in the Pacific Northwest?

Driving up Marys Peak you're right above the eastern edge of the "locked zone" between the colliding Juan de Fuca and North American plates. The plates lock together for centuries until they just can't take it any more, releasing suddenly and violently. Research over the past four decades has revealed that giant earthquakes have occurred periodically on this zone, and that others will surely happen. The last great earthquake occurred just over 300 years ago, on January 26, 1700. When the plates suddenly unlock again, the resulting earthquake likely will be devastating. Most of the people in Oregon live in the Willamette Valley, just inland from the locked zone. Compared to the hard, durable rock layers of the Coast Range, the soft sedimentary layers of the Willamette Valley could shake like a bowlful of jelly during the earthquake. That, coupled with landslides and tsunamis (large sea waves) generated by the earthquake, would devastate coastal regions and cut them off from food, medicine, and other supplies found in the more populated regions of the Willamette Valley.

6. What unique opportunities does the top of Marys Peak provide?

The summit of Marys Peak is a beautiful and inspiring place to observe many of the natural and cultural aspects of the surrounding landscape. It is one of the few places on Earth where one can observe the full surface expression of an active subduction zone—from the blue waters of the Pacific Ocean (subducting Juan de Fuca Plate), to the rolling hills of the Coast Range (accretionary wedge), to the lush green Willamette Valley (forearc basin), to the snow-capped peaks of the Cascade Range (volcanic arc). It is also a place to contemplate the effects of the Missoula Floods, which filled the Willamette Valley with 300 feet of water when ice dams on the edge of the Canadian Ice Sheet periodically broke 16,000 to 12,000 years ago. Enormous volumes of water roared down the Columbia River and up into the Willamette Valley.

The view from the top of the peak provides an opportunity to observe various ecosystems of western Oregon and to see the effects of agricultural and forestry practices on the landscape. You can also see other mountains that, like Marys Peak, have been sacred places for vision quests by the original Kalapuya inhabitants of the region.

Marys Peak Geology Road Log

The summit of Marys Peak is at the end of Marys Peak Road, accessed from Oregon Highway 34 just east of the pass over the Coast Range known as Alsea Summit. Traveling westward through Philomath, you come to what is locally called the "Y-Intersection" of U. S. Highway 20 and Oregon Highway 34. Turn left on Highway 34 toward Alsea and Waldport. The right-hand turnoff onto Marys Peak Road is 8.9 miles from the Y-Intersection. From the west travel through Alsea on Highway 34 and turn left after 7.9 miles, just beyond Alsea Summit.

Stops 1 and 6 are on Highway 34 east and west of Marys Peak Road, respectively. The highway can be busy and cars and trucks travel at highway speeds, so please use extreme caution. These roadcuts can be viewed most safely by parking and remaining on the south side of the road, opposite the roadcuts. Stops 2 through 5 are along Marys Peak Road. Although traffic does not generally travel as fast there as along Hwy 34, caution should still be used in parking and walking to roadcuts.

The road guide is tied to a series of geological maps and a simplified geological cross section of the Marys Peak region. The basic layering of Oregon's central Coast Range includes oceanic lava flows of basalt overlain by sedimentary layers of sandstone and shale. Later intrusions of hot magma solidified as very hard igneous rock known as gabbro. The layers were folded into an anticline as they were pushed upward and eastward along the Corvallis Fault. Erosion removed up to two miles of rock from the top of the anticline, exposing the layers seen on a drive up Marys Peak.

Stops 1 and 3 reveal the sedimentary layers in steeply tilted and nearly flat orientations, respectively. Stop 2 shows dramatic examples of pillow structures formed as lava flowed and cooled on the ocean floor. The hard, resistant gabbro is seen forming a waterfall at Stop 4 and the summit of Marys Peak at Stop 5. Stop 6 shows a dike of gabbro that represents a vertical conduit of magma that fed thick sills of gabbro such as those capping Marys Peak and other Coast Range mountains.

Detailed Geology and Road Log Stops

Marys Peak is on a gabbro intrusion into Siletz River Volcanics and marine sedimentary rocks. The road trip visits tilted sedimentary layers (Stop 1), basalt that erupted beneath the ocean (Stop 2), flat-lying sedimentary layers (Stop 3), gabbro that intruded as a sill (parallel to the other rock layers, Stops 4 and 5), and gabbro that intruded as a dike (cutting across the other layers, Stop 6). The parking lot or short walk to the summit at Stop 5 provides one of the world's best views of the landscape of an active subduction zone.

Marys Peak Geology Road Log		
Stops and Intersections	**Mileage* Marys Peak Rd** (Y-Intersection)	**Geological Features and Notes**
Hwy 20/34 Intersection in Philomath	(0.0)	Locals refer to this as the "Y-Intersection." Take Hwy 34 southwest toward Alsea and Waldport. *GPS: 44.5413° N, 123.3853° W*
Stop 1	(7.4)	**Tilted Sandstone and Shale Layers**. (Tyee Formation). Park in gravel area on inside of switchback curve. **(Caution: Busy Highway)** *GPS: 44.4716° N, 123.4841° W*
Intersection: Hwy 34 and Marys Peak Rd	**0.0** (8.9)	Turn right (north) on Marys Peak Road. *GPS: 44.4672° N, 123.5033° W*
Stop 2	**3.7** (12.6)	**Pillow Basalt (Siletz River Volcanics)**. Park in large parking lot on left. Facing away from road, walk ¼ mile up gravel road at far left corner of lot. *GPS: 44.4815° N, 123.5354° W*
Stop 3	**6.4** (15.3)	**Flat-lying Sandstone and Shale Layers**. (Tyee Formation). Just beyond layers, pull out to left as far as possible. Walk back to flat layers. **(Caution on Road)** *GPS: 44.5003° N, 123.5605° W*
Stop 4	**6.8** (15.7)	**Gabbro Sill**. Parking area on right, at Parker Creek Falls. *GPS: 44.5041° N, 123.5630° W*
Stop 5	**9.5** (18.4)	**Panorama of Cascadia Subduction Zone**. Park in lot at end of paved road (U.S. Forest Service permit required). Walk ½ mile up gravel road to the top of Marys Peak. *GPS: 44.5104° N, 123.5450° W*
Intersection: Marys Peak Rd and Hwy 34	**0.0** (8.9)	Drive back to the beginning of Marys Peak Road and turn right (west) on Hwy 34 toward Alsea. *GPS: 44.4672° N, 123.5033° W*
Stop 6	(9.0)	**Gabbro Dike**. Drive past Alsea Summit and park in gravel area at orange gate on left (south) side of road. Walk back up Hwy 34 to observe dike in road cut on north side of road. **(Caution: Busy Highway)** *GPS: 44.4647° N, 123.5069° W*

*The **bold** numbers correspond to the brown mileage markers along Marys Peak Road.
Numbers in (parentheses) are distances along Highway 34 from the Y-Intersection in Philomath.

Marys Peak Cross Section and Road Stops

Stop 2
Pillow Basalt

Rock quary off Marys Peak Road, right before the saddle

Stop 1
Sandstone and Shale (Tilted)

Switchbacks on Hwy 34, 1.5 miles northeast of Marys Peak Road

Stop 4
Gabbro (Sill)

Parker Creek Falls

Stop 3
Sandstone and Shale (Flat)

Marys Peak Road, past turnoff to Harlan

Stop 6
Gabbro (Dike)

Alsea Summit on Hwy 34, just west of turnoff to Marys Peak Road

Stop 5
Gabbro (Sill)

Walking up the gravel road, just before the top of Marys Peak

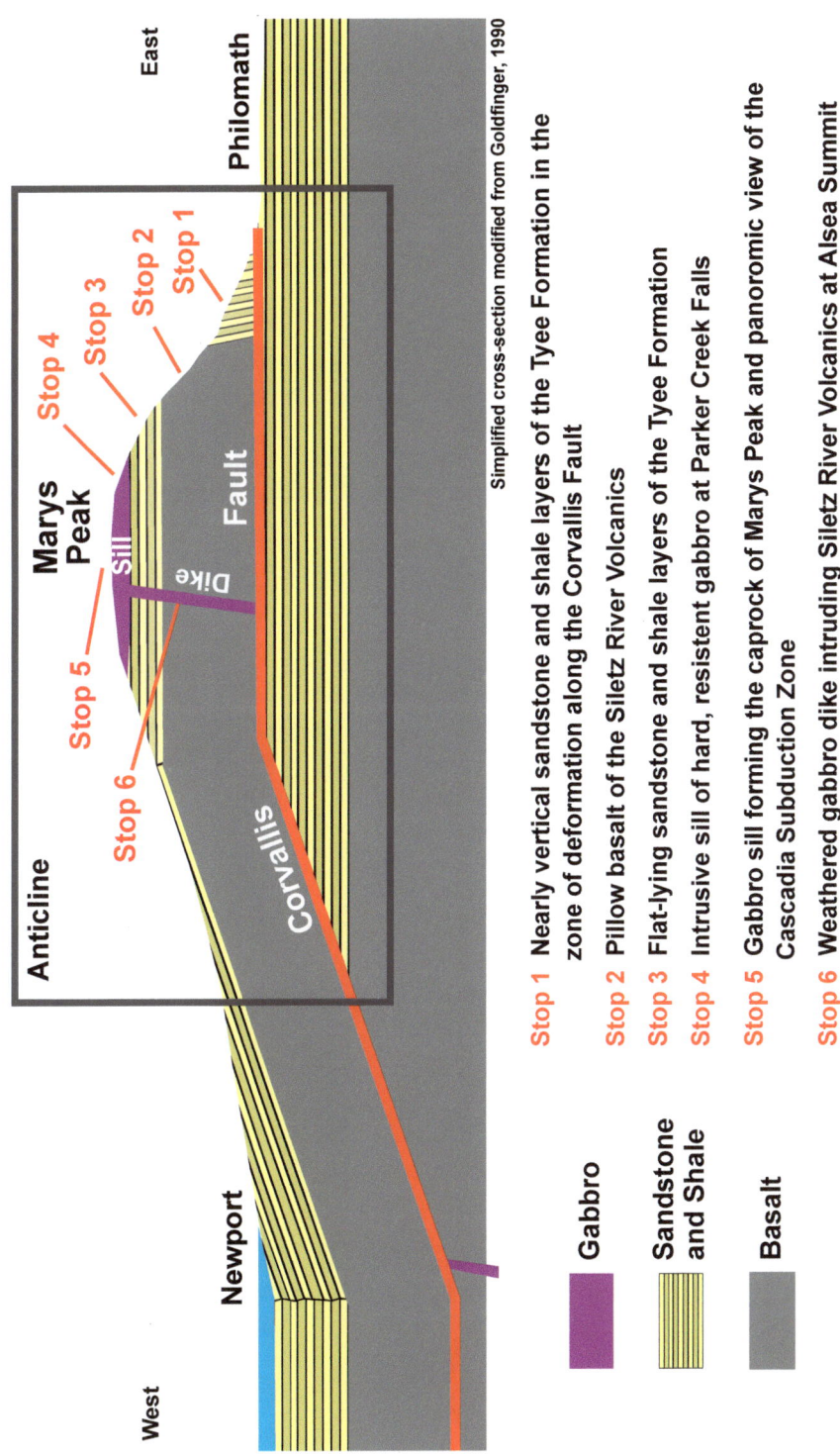

Simplified cross-section modified from Goldfinger, 1990

Stop 1   Nearly vertical sandstone and shale layers of the Tyee Formation in the zone of deformation along the Corvallis Fault

Stop 2   Pillow basalt of the Siletz River Volcanics

Stop 3   Flat-lying sandstone and shale layers of the Tyee Formation

Stop 4   Intrusive sill of hard, resistent gabbro at Parker Creek Falls

Stop 5   Gabbro sill forming the caprock of Marys Peak and panoromic view of the Cascadia Subduction Zone

Stop 6   Weathered gabbro dike intruding Siletz River Volcanics at Alsea Summit

Gabbro

Sandstone and Shale

Basalt

11

Chapter 1

Landscape of Marys Peak and the Pacific Northwest
From Ancient Seafloor to Modern Mountains

Marys Peak is one of the few places on Earth where, on a clear day, you can see all the surface elements of an active subduction zone and contemplate forces beneath the surface that result in earthquakes and volcanic eruptions. As one of the premier outdoor destinations in the Pacific Northwest, it is an amazing place to observe geologic processes in action. The Coast Range of Washington, Oregon and northern California contains rocks that were manufactured in the ocean, then scraped off the subducting Juan de Fuca plate and lifted out of the sea. Farther inland, the Cascade Mountains are dramatically different. They are explosive volcanoes forming as fluids rise from the top of the subducting plate and generate magma that makes its way to the surface. And as time ticks on, the region awaits sudden release of energy locked between the converging plates as a devastating earthquake.

Plate Tectonics

The landscapes of the Pacific Northwest, as well as geologic hazards such as earthquakes and volcanic eruptions, are due to the movement of large plates of Earth's outer shell. There are three types of tectonic plate boundaries. Plates rip apart at a *divergent boundary*, causing volcanic activity and shal-

← Sandstone and shale layers are hardened remnants of sand and mud deposited on an ancient seafloor and lifted high above sea level on Marys Peak.

Most Earthquakes and Volcanic Eruptions Occur at Plate Boundaries and Hotspots

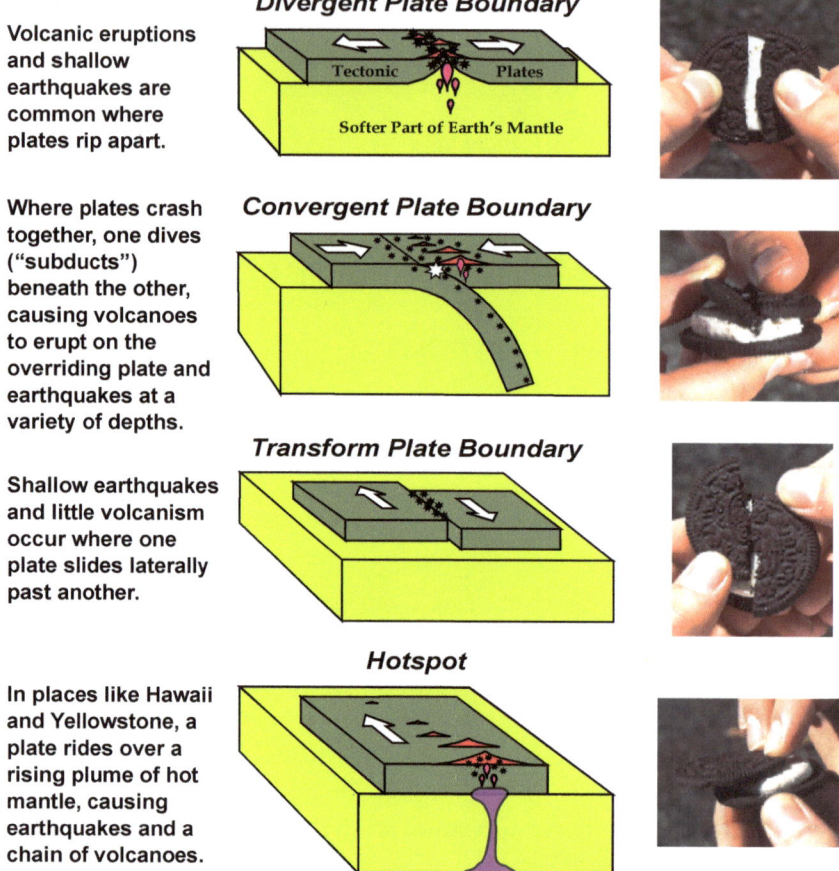

Divergent Plate Boundary

Volcanic eruptions and shallow earthquakes are common where plates rip apart.

Tectonic   Plates

Softer Part of Earth's Mantle

Convergent Plate Boundary

Where plates crash together, one dives ("subducts") beneath the other, causing volcanoes to erupt on the overriding plate and earthquakes at a variety of depths.

Transform Plate Boundary

Shallow earthquakes and little volcanism occur where one plate slides laterally past another.

Hotspot

In places like Hawaii and Yellowstone, a plate rides over a rising plume of hot mantle, causing earthquakes and a chain of volcanoes.

Volcanoes ▲ Earthquakes ✴ Small to Moderate Size
☆ Very Large

Oreo® cookies are a fun way to demonstrate the three types of plate boundaries and a hotspot☺.

low earthquakes. At a *convergent boundary*, one plate dives ("subducts") beneath the other, resulting in a variety of earthquakes and a line of volcanoes on the overriding plate. *Transform boundaries* are where plates slide laterally past one another, producing shallow earthquakes but little or no volcanic activity. Another large-scale feature is a *hotspot*, where a plate rides over a rising plume of hot mantle, creating a line of volcanoes on the top of the plate. The greater Pacific Northwest is one of the few places on Earth that has active examples of all three types of plate boundaries and a hotspot.

The Greater Pacific Northwest Has all Three Types of Plate Boundaries and a Hotspot

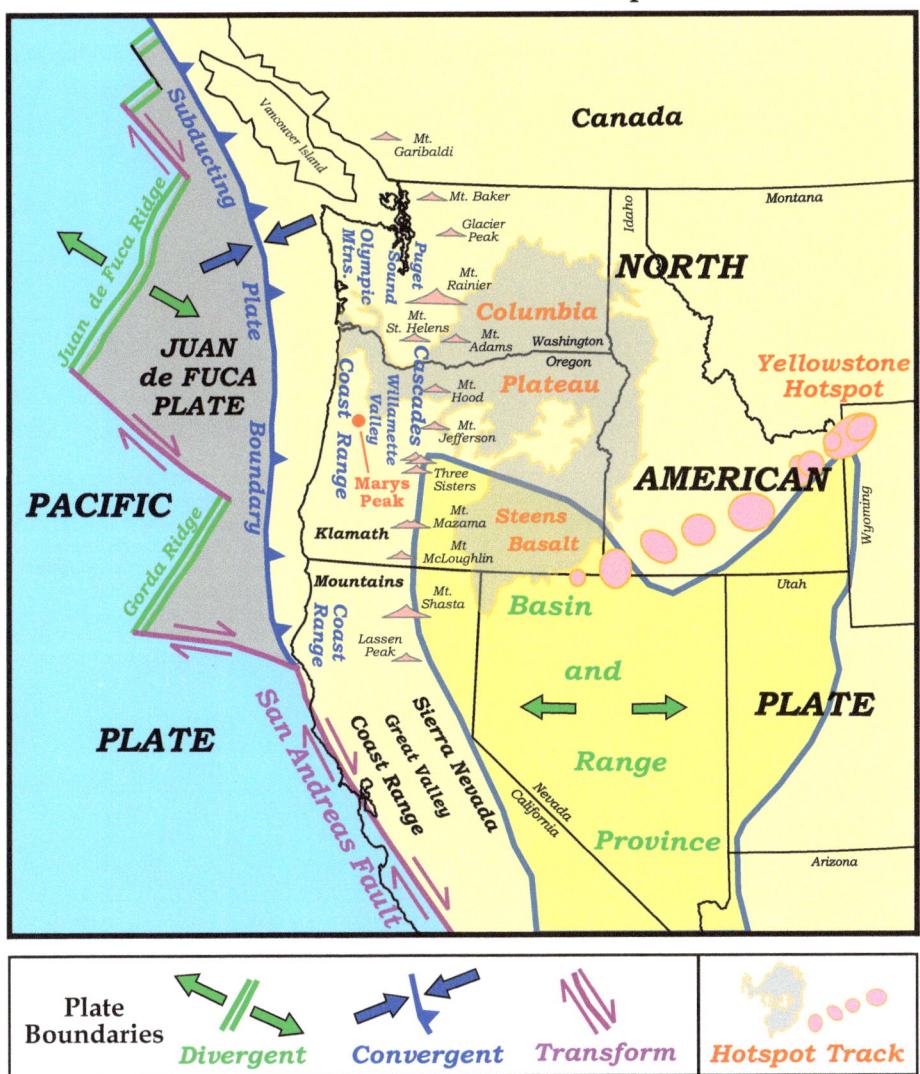

Subduction of the Juan de Fuca Plate beneath North America forms the coastal ranges and Cascade volcanoes. Offshore, the Juan de Fuca Plate diverges from the Pacific Plate at the Juan de Fuca and Gorda ridges. Between the ridge segments, the plates slide past one another along transform plate boundaries similar to California's San Andreas Fault. Onshore, the Basin and Range Province forms long mountain ranges and intervening valleys as plate divergence rips North America apart. The Columbia Plateau and Steens Basalt represent the surfacing of a hotspot now located beneath Yellowstone National Park.

Plate Convergence Leads to Subduction and the Formation of two Parallel Mountain Ranges

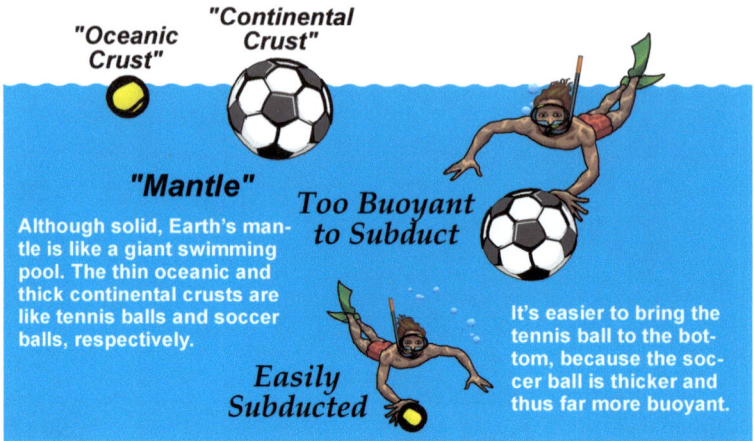

"Oceanic Crust"

"Continental Crust"

"Mantle"
Although solid, Earth's mantle is like a giant swimming pool. The thin oceanic and thick continental crusts are like tennis balls and soccer balls, respectively.

Too Buoyant to Subduct

Easily Subducted

It's easier to bring the tennis ball to the bottom, because the soccer ball is thicker and thus far more buoyant.

Where tectonic plates converge, the one with thin oceanic crust subducts beneath the one with thick, more buoyant continental crust.

Near their boundary, the plates can lock together for centuries, then suddenly let go as a giant earthquake. If the seafloor rises or falls, giant sea waves (a tsunami) can form.

An accretionary wedge forms between the converging plates as material is scraped off the subducting plate.

Farther inland, the subducting plate reaches depths where it "sweats" hot water. The rising water melts rock in its path, forming a volcanic arc on the overrriding plate.

A forearc basin develops in the low area between the two mountain ranges.

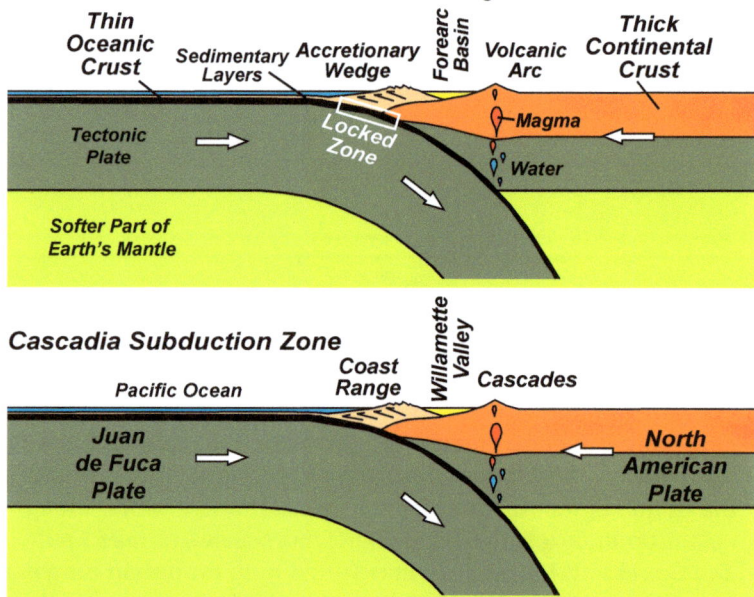

The Coast Range and Cascades are the two parallel mountain ranges in the Pacific Northwest. The forearc basin is the Willamette Valley in Oregon and Puget Sound in Washington.

The Cascadia Subduction Zone has two Parallel Mountain Ranges Separated by a low Region near Sea Level

The **COASTAL MOUNTAIN RANGES**, including the Olympic Mountains in northwest Washington and the Coast Range in southwest Washington, western Oregon and northwest California, form as sedimentary and volcanic layers are scraped off the top of the subducting oceanic plate and added to the edge of the continent.

The low region between the two parallel mountain ranges is the **PUGET SOUND** area in Washington, the **WILLAMETTE VALLEY** in Oregon and the **GREAT VALLEY** in northern California.

150 miles inland, the top of the subducting plate reaches depths where it's hot enough to generate fluids, forming volcanoes in the **CASCADES**.

National Park Service

U. S. Forest Service

NP = National Park
NM = National Monument
NVM = National Volcanic Monument
NHP = National Historical Park
NRA = National Recreation Area
N&SP = National and State Parks
RA = Recreation Area
SA = Scenic Area

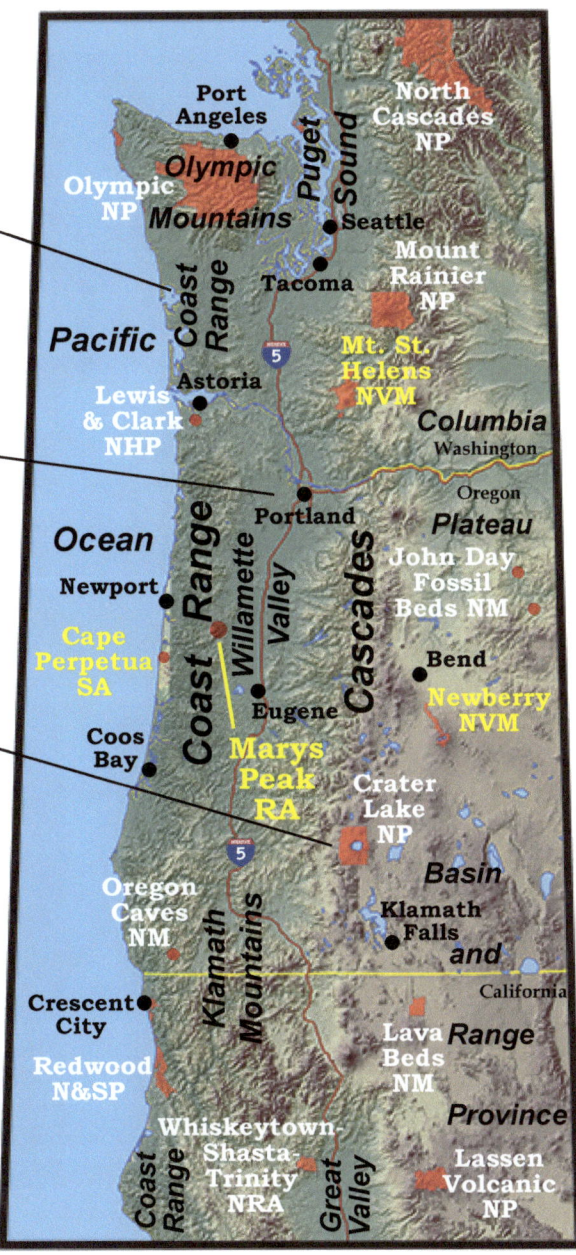

Two Parallel Mountain Ranges Form along the Cascadia Subduction Zone

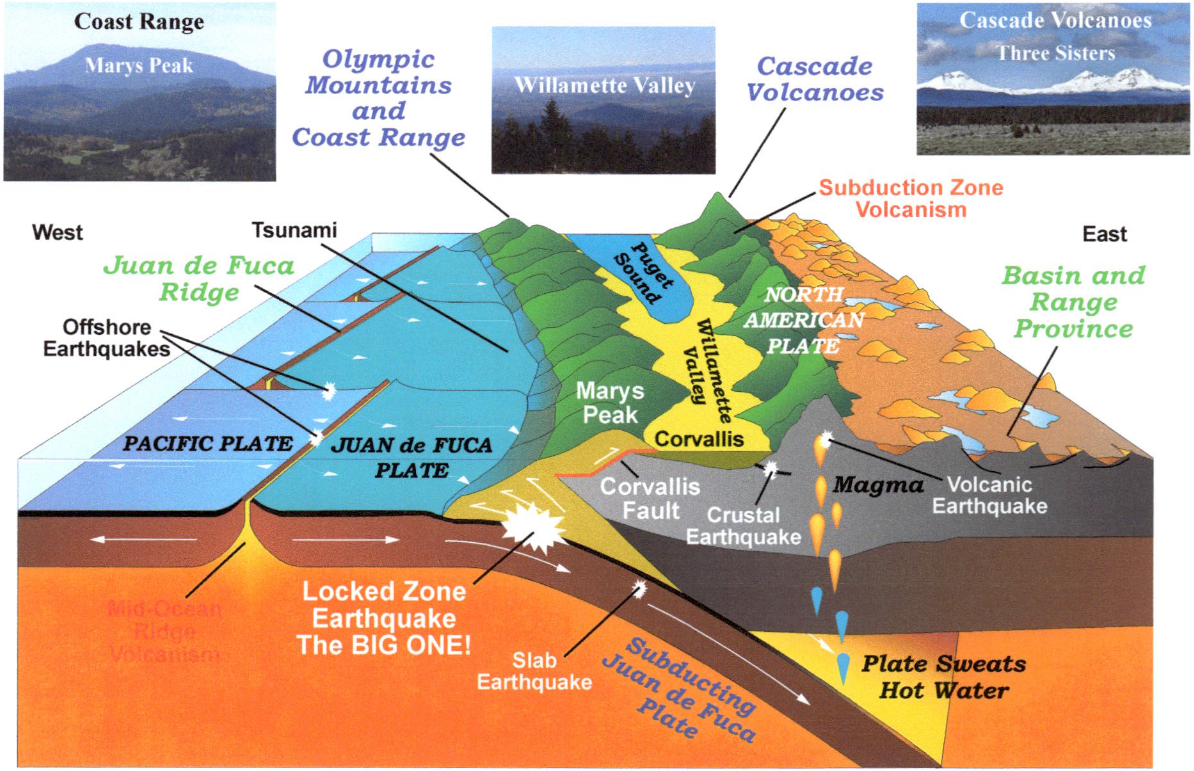

Coast Range
Marys Peak

Willamette Valley

Cascade Volcanoes
Three Sisters

West

East

Olympic Mountains and Coast Range

Cascade Volcanoes

Subduction Zone Volcanism

Basin and Range Province

NORTH AMERICAN PLATE

Willamette Valley

Puget Sound

Marys Peak

Corvallis

Corvallis Fault

Crustal Earthquake

Magma Volcanic Earthquake

Plate Sweats Hot Water

Subducting Juan de Fuca Plate

Slab Earthquake

Locked Zone Earthquake
The BIG ONE!

Juan de Fuca Ridge

JUAN de FUCA PLATE

PACIFIC PLATE

Tsunami

Offshore Earthquakes

Mid-Ocean Ridge Volcanism

The Coast Range consists of igneous and sedimentary rock layers formed in the ocean and scraped off the subducting Juan de Fuca Plate. Marys Peak is part of a block of Coast Range rocks that were lifted up and thrusted over the edge of the Willamette Valley along the Corvallis Fault. The Cascade Volcanoes develop farther inland, where the plate is subjected to so much heat and pressure at depth that it dehydrates ("sweats"), causing overlying rock to melt.

The ***Basin and Range Province*** is a continental rift zone that includes all of Nevada and adjacent portions of Oregon, Idaho and California. It is a divergent plate boundary where the North American Plate is ripping apart, forming long mountains (ranges) separated by valleys (basins). Another type of divergent plate boundary occurs one to two hundred miles off the coast of Washington, Oregon and northern California, where the Pacific Ocean floor is ripping apart along the Juan de Fuca and Gorda ridges. These small mid-ocean ridges are similar to the Mid-Atlantic Ridge forming where the Atlantic Ocean is opening due to divergence of the Eurasian and African plates away from the North and South American plates.

The world's most famous transform plate boundary, the ***San Andreas Fault***, transports a sliver of western California northward where the Pacific Plate slides laterally past the North American Plate. Other transform plate boundaries occur off the coast of the Pacific Northwest, connecting the northern end of the San Andreas Fault to the Gorda Ridge, and the Gorda Ridge to the Juan de Fuca Ridge. These are zones where the Juan de Fuca Plate slides laterally past the Pacific Plate.

The Hawaiian Islands have formed progressively over the past 5 million years as the Pacific Plate moves northwestward over the Hawaiian Hotspot. A similar volcanic track has been forming over the

Forming the Coast Range

Jennifer Natoli

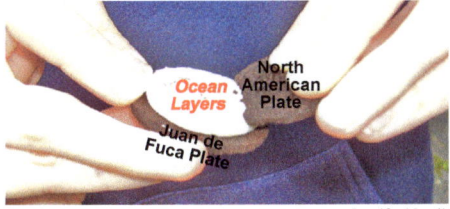

Jennifer Natoli

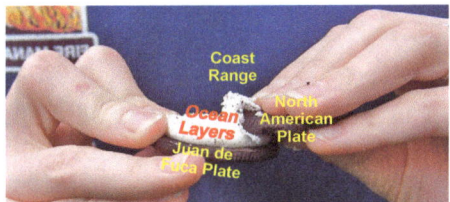

Jennifer Natoli

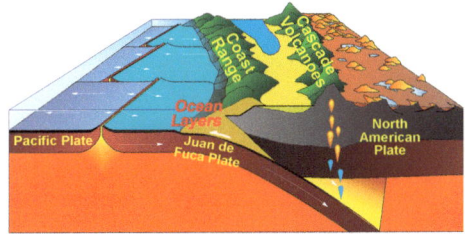

Ranger Jen's Oreo Demo. Jennifer Natoli was a seasonal ranger at Redwood National and State Parks in California. In her version of the Oreo® cookie demonstration, the creamy filling is the layers of sediment and basalt on the ocean floor. As the Juan de Fuca Plate (lower cookie) subducts beneath the North American Plate (upper cookie), the layers are scraped off the ocean floor and pile up as the Coast Range.

past 17 million years in the Pacific Northwest as the North American Plate has drifted in a west-southwest direction over the *Yellowstone Hotspot.* The hotspot initially reached Earth's surface 17 million years ago, resulting in eruption of the massive Steens Basalt and Columbia Plateau Basalt in eastern Oregon and Washington. A track of supervolcanoes then formed across southern Idaho as the plate drifted over the hotspot. Yellowstone National Park, with its spectacular geysers, hot springs and other volcanic and geothermal features, is currently directly above the hotspot.

The *Cascadia Subduction Zone*, extending from northern California through western Oregon and Washington to southern British Columbia, is a convergent plate boundary. As the Juan de Fuca Plate moves eastward, it carries hard crust and sediment layers toward North America, as if on a giant conveyor belt. At the edge of the continent, the plate plunges downward and some of the layers are scraped off the top and squeezed upward as the Olympic Mountains and other coastal ranges. Farther east the top of the Juan de Fuca Plate descends deeper and deeper. The sedimentary layers and hard crust are metamorphosed due to the great temperatures and pressures at those depths. A by-product of the metamorphism is the release of hot fluids, especially water. The water rises and melts rock in its path. Some of the resulting magma makes it all the way to the surface and forms Mount Rainier, Mount Hood, Mount Shasta and other spectacular volcanoes of the Cascade Range.

Earthquakes and Volcanic Eruptions

A drive up Marys Peak is an opportunity to appreciate the "beauty from the beast" aspect of Oregon's landscape. It reveals how the same geological forces that threathen our lives with earthquakes and volcanic eruptions also nourish our spirits by forming the spectacular mountains, valleys and coastlines of the Pacific Northwest. Individual earthquakes can offset or lift the land a fraction of an inch or a few inches at a time. That may not seem like much, but when tens of thousands of earthquakes occur over a few million years, coastal ranges form. Likewise, individual volcanic eruptions may add only a thin layer to the surface. But thousands of eruptions over less than a million years have built Mount Hood and other Cascade peaks to elevations of more than 10,000 feet!

Earthquakes

A variety of earthquakes shake the Pacific Northwest due to plate-tectonic activity (diagram, page 18). The largest (*locked zone earthquakes*) occur where the Juan de Fuca and North American plates are stuck together, as they have been for the past three centuries. When the plates suddenly let go, a massive earthquake will shake the entire Pacific Northwest, a series of tsunami waves will pound the Coast, and landslides will make it difficult to reach some of those in need. These mega-earthquakes occur every 200 to 600 years or so, and the last one was in the year 1700. We are wise to prepare our homes, communities, and infrastructure for the next "Big One," and to know what to do during the ground shaking, landslides, and tsunamis that are sure to come.

The moving plates can cause destruction in other ways. *Slab earthquakes*, like the magnitude 6.8 Nisqually Earthquake that struck Washington's Puget Sound region in 2001, occur within the subducting Juan de Fuca Plate, as it bends and breaks. Populations in the Portland and Seattle areas are also at risk from *crustal earthquakes* that occur on shallow faultlines where the crust of North America snaps in the vise between the two converging plates. An example is the 1993 magnitude 5.6 Scotts Mills Earthquake, which occurred on the Mount Angel Fault beneath the eastern Willamette Valley. The Corvallis Fault is a similar crustal fault, but it is not clear when movement last occurred along that fault line and if it poses a risk for future earthquake shaking. And as magma moves to shallow depths beneath the Cascades, small earthquakes and tremors occur. Although these *volcanic earthquakes* are generally not particularly damaging, they are sometimes clues to impending eruptions, such as the one that occurred on Mount St. Helens in 1980.

Coastal residents often awake to reports of small to moderate-sized earthquakes. These *offshore earthquakes* are usually not along the subducting plate boundary. Rather, they occur at the divergent and transform plate boundaries between the Pacific and Juan de Fuca plates. These earthquakes are not likely to cause destruction because they are too small and too far away, and because they do not cause significant up or down movement of the seafloor needed to generate a tsunami.

Knowledge of these earthquakes—and what to do to prepare for them—comes not only from scientific studies over the past century, but also from indigenous people who passed along knowledge and wisdom from generation to generation. The most recent great earthquake on the Cascadia Subduction Zone occurred in the year 1700. There were no written records kept in the Pacific Northwest at that time to document the event. But geologists are detectives. Using not only scientific observations and analyses, but also historical and anthropological evidence, they were able to piece together clues that unraveled the mystery. These clues suggest that there is about a one-in-three chance that the next great Cascadia Subduction Zone

Coast Range and Pacific Ocean

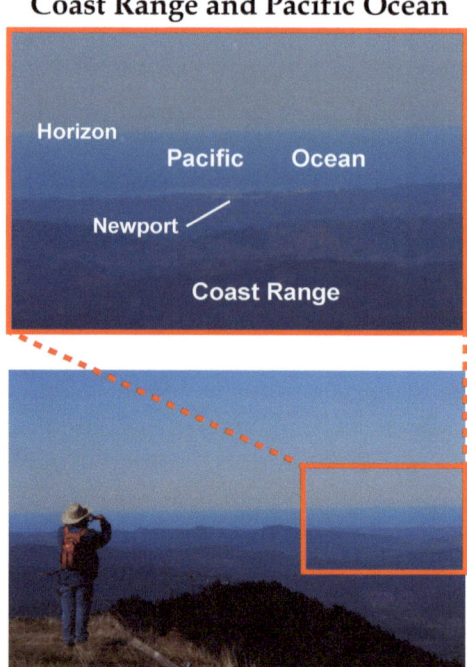

On a clear day you can see the Pacific Ocean from Marys Peak. The coastline at Newport is about 40 miles away. The horizon of the ocean roughly represents where the Juan de Fuca Plate begins its descent beneath the edge of North America.

earthquake—and accompanying tsunami—will strike the Pacific Northwest in the next 50 years.

Volcanic Eruptions

The top of Marys Peak provides an exceptional setting to appreciate the danger and likelihood of major volcanic eruptions in the Pacific Northwest. From that magnificent viewpoint you can imagine the Juan de Fuca Plate beginning its downward plunge about 50 miles out in the Pacific Ocean. The plate encounters increasing temperatures and pressures as it extends deep into the Earth. Beneath the Cascade Mountains, the top of the plate is about 50 miles deep, where the heat and pressure cause rock to metamorphose—that is, its minerals recrystallize to other forms. A by-product of the metamorphism is the release of fluids, particularly water. In other words, just like us, when the plate is under a lot of pressure and can't stand the heat, it sweats! Some of the water rises into the overriding plate (in this case, the North American Plate). When the rising water encounters the already hot

Willamette Valley and Cascades

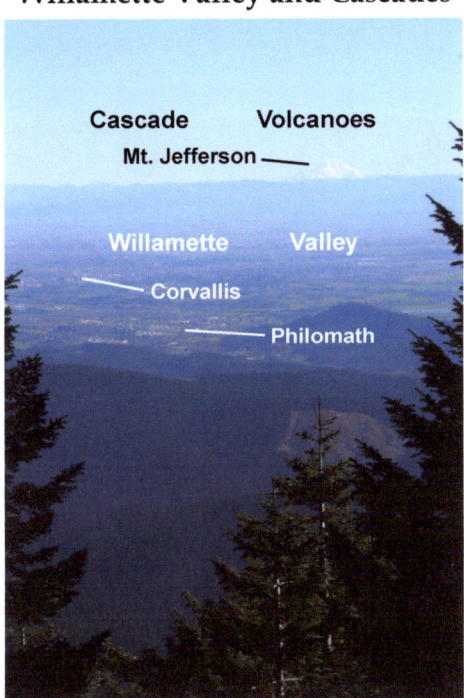

On a clear day you can see high Cascade peaks from Marys Peak. The volcanoes represent the line where the top of the Juan de Fuca Plate reaches about 50 miles depth.

mantle rock of that plate, it causes minerals in the rock to melt. Some of the resulting magma makes it all the way to the surface and erupts, forming the Cascade Volcanoes. The line of volcanoes extends all the way from Mount Garibaldi in southern British Columbia to Lassen Peak in northern California. On a very clear day a 350-mile swath of this line can be seen from the top of Marys Peak, from Mount Rainier in Washington to Mount McLoughlin in southernmost Oregon.

Other volcanic activity occurs off the coast of Oregon and Washington, but this activity is not due to the Cascadia Subduction Zone. Rather, the hot mantle begins to melt as it shallows beneath the Juan de Fuca and Gorda ridges, causing basalt lava to erupt on the ocean floor. Similar but ancient basalt lava flows are seen in spectacular fashion at Stop 2 of this Road Guide, where their pillow forms indicate that the lava flowed and solidified beneath water rather than on land. The fact that this basalt—which may have formed up to three miles beneath the ocean—is now a half-mile above sea level, reveals that dynamic geological forces have acted on the rock layers of Marys Peak.

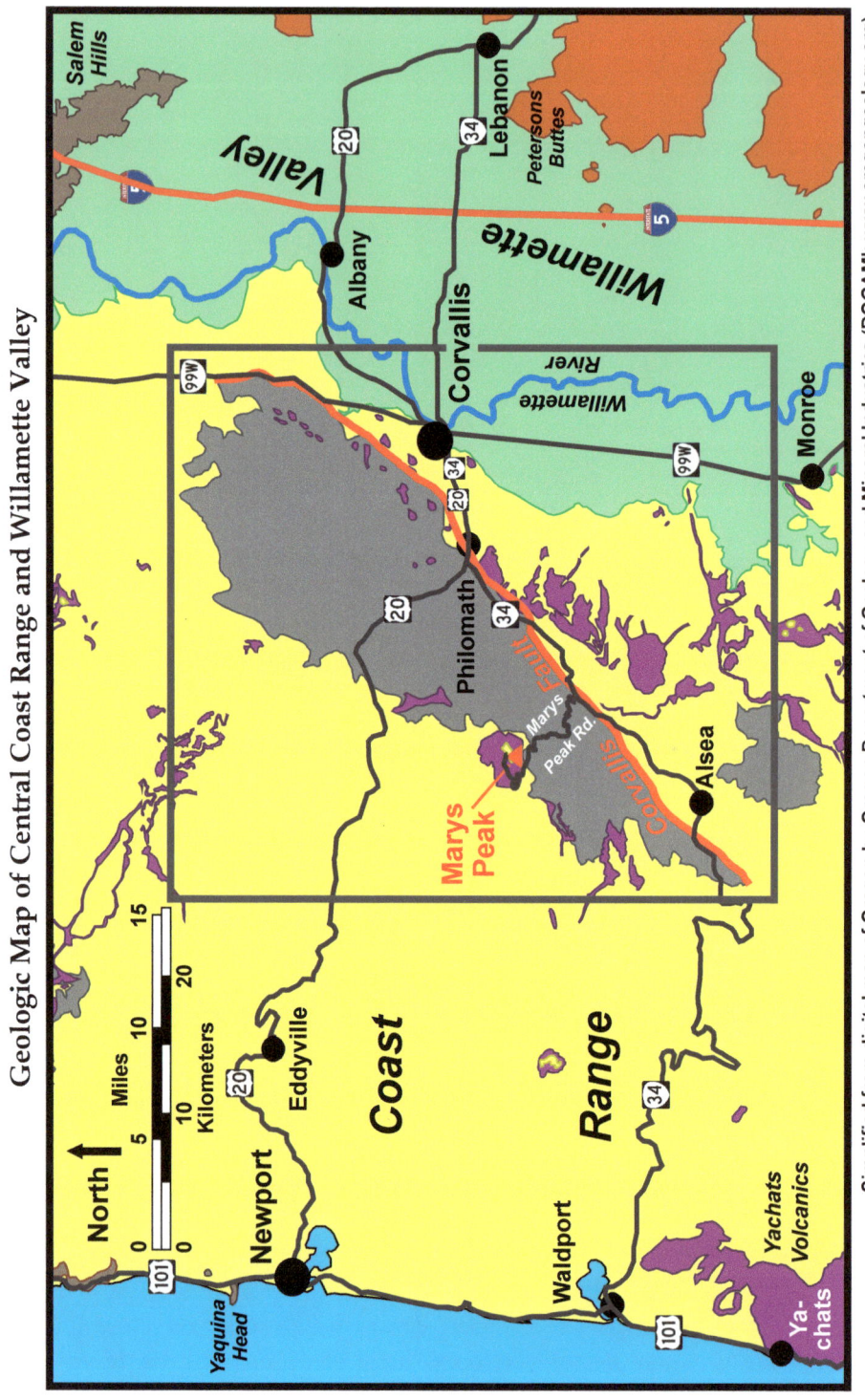

Geologic Map of Central Coast Range and Willamette Valley

Simplified from digital map of Oregon by Oregon Department of Geology and Mineral Industries (DOGAMI; www.oregongeology.org).

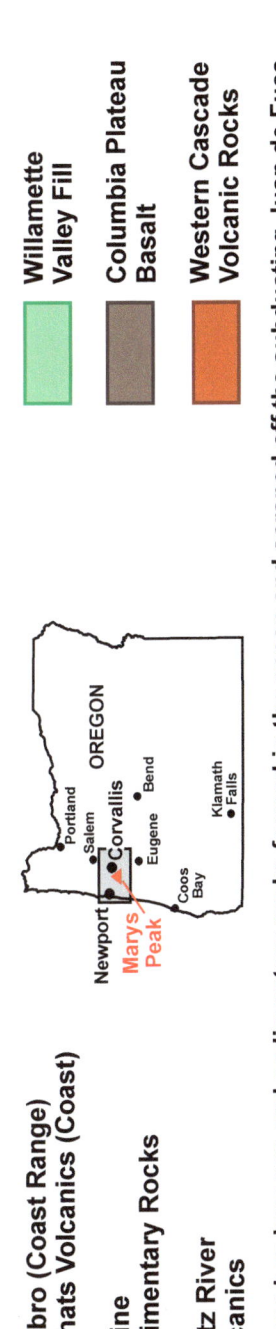

Willamette Valley Fill

Columbia Plateau Basalt

Western Cascade Volcanic Rocks

Gabbro (Coast Range) Yachats Volcanics (Coast)

Marine Sedimentary Rocks

Siletz River Volcanics

The Coast Range has igneous and sedimentary rocks formed in the ocean and scraped off the subducting Juan de Fuca Plate. The Marys Peak area is part of a block of Siletz River Volcanics that was lifted up and thrusted over the edge of the Willamette Valley along the Corvallis Fault. Young intrusions of gabbro form the caprock of many Coast Range mountains, including Marys Peak. Extrusive lavas of the same age poured out on the surface as the Yachats Volcanics (southwest corner of map). Columbia Plateau Basalt lavas flowed all the way from eastern Oregon to the Salem Hills (northeast corner of map) and Pacific Coast (northwest corner of map). Sediment filling the Willamette Valley includes lake and stream deposits and layers from catastrophic ice age floods. Box on the large map shows area of map in Chapter 2.

OREGON

Portland
Salem
Corvallis
Bend
Newport
Eugene
Marys Peak
Coos Bay
Klamath Falls

Regional Geologic Map

The map of Oregon's central Coast Range and adjacent Willamette Valley shows the surface exposure of rocks formed due to Cascadia Subduction Zone processes—and even due to surfacing of the Yellowstone Hotspot. Coast Range rocks include basalt and overlying sedimentary layers that were deposited on the seafloor far out in the Pacific Ocean. The basalt, known as the Siletz River Volcanics, is part of a series of similar-age rocks extending along the length of the coastal ranges. These basalt outpourings may have formed in a variety of ways, including along mid-ocean ridges or as parts of volcanic island chains like Hawaii. After being covered by sand and mud eroded from the North American continent, these "terranes" slammed into the coast of the Pacific Northwest. Locally the block of Siletz River Volcanics is known as "Siletzia," a terrane that crashed into western Oregon and was lifted upward and pushed eastward along the Corvallis Fault. The uplift was accompanied by erosion from wind, water and ice, exposing an anticline that has older rock (basalt) in its center, surrounded by younger sandstone and shale layers.

Prior to uplift, the basalt and its sedimentary cover were intruded by magma with chemistry similar to basalt. In most places it cooled slowly beneath the surface, forming a series of dikes (cutting across layers) and sills (inserted parallel to layers). The slow cooling produced coarse-grained, dark-colored rock known as gabbro. This rock is the caprock of Marys Peak and many of the other high mountains in Oregon's central Coast Range. Some of the magma did make it to the surface and is exposed along the Coast as the Yachats Volcanics.

There are volcanic rocks of two other ages that appear on the regional map. Eruptions on the eastern side of the Willamette Valley formed Petersons Buttes near Lebanon and other early-Cascade volcanic features. Younger lavas flowed from the Columbia Plateau of eastern Oregon and Washington, through the Columbia Gorge and all the way to the Willamette Valley and Pacific Coast. The Salem Hills in the northeastern corner of the map include some of these hard and resistant lava flows, including those that form the spectacular waterfalls of Silver Falls State Park. Many of the resistant headlands on the north and central Oregon Coast, including Yaquina Head just north of Newport, are made of Columbia Plateau Basalt.

The Willamette Valley is filled with young lake and river deposits that cover the older rock layers. It also has mud, sand and gravel—and even large boulders—that were brought in by enormous floods during the waning stages of the last ice age. Known as the Missoula Floods, these floods occurred when ice dams forming a giant lake in Montana (Lake Missoula), periodically gave way and sent incredible volumes of water down the Columbia River. The flood waters were deflected southward up the Willamette Valley, periodically forming a turbulent lake as much as 300 feet deep in the Corvallis area. The boulders were not carried directly by the floodwaters, but rather were embedded in icebergs that were left behind and melted. One can only imagine what the early inhabitants of the Willamette Valley—the Kalapuya—might have witnessed while watching the floodwaters and their aftermath from the high summit of Marys Peak.

Chapter 2

Forming Marys Peak
The Corvallis Fault and
Stubborn Igneous Rock

Marys Peak is an "island in the sky" for three reasons. First, it is part of the Coast Range that formed as layers were scraped off the ocean floor during subduction. Second, it is within a block of oceanic material (basalt capped by sandstone and shale) that was shoved upward and eastward along the Corvallis Fault. And third, its upper portions have an intrusive rock layer (gabbro) that is thick and very hard, so that it resists erosion.

Marys Peak has volcanic rocks, but Marys Peak is *not* a volcano. So how did the ancient geological layering and present-day structure of Marys Peak come to be?

The Marys Peak story begins with the eruption of basalt lava flows on the floor of the Pacific Ocean about 55 million years ago. Sand and mud layers were deposited on top of the basalt for the next 20 million years as the oceanic plate moved toward the edge of North America. The block containing the basalt and its sedimentary cover slammed into the edge of the continent about 35 million years ago. At the same time, coarse-grained igneous rock intruded into the basalt and sedimentary layering. Through continued plate convergence, the whole block was squeezed upward and thrust eastward over the edge of the Willamette Valley along the Corvallis Fault. Rock layers were deformed and tilted to steep angles near the fault line, but remained relatively flat near the top. Simultaneous erosion removed much of the sedimentary and volcanic layering, but the highly resistant igneous

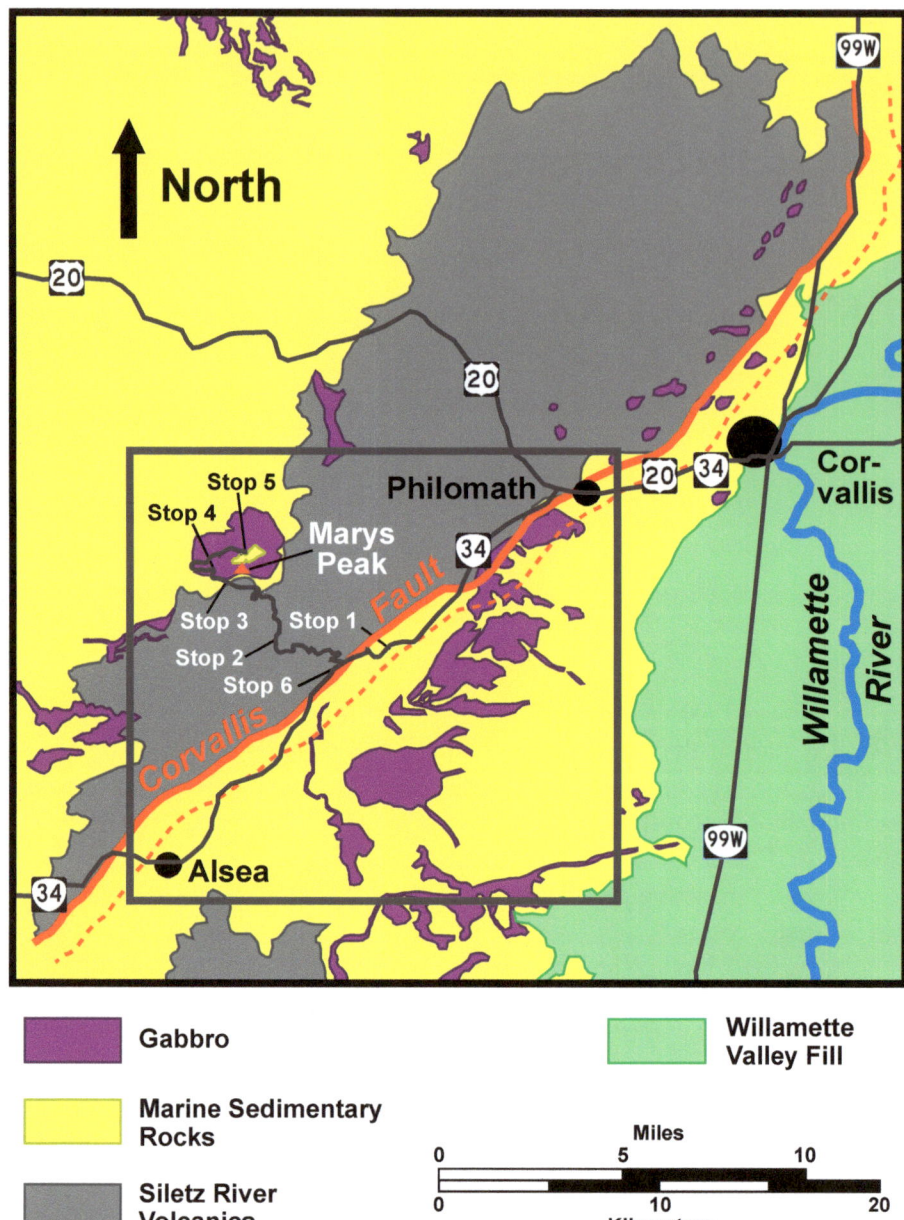

Marys Peak is the top part of a block of basalt (Siletz River Volcanics) and its sedimentary cover that was uplifted and thrust eastward along the Corvallis Fault, forming a large upfold (anticline). The dashed red line shows the approximate extent of a zone of deformed and steeply tilted layers on the eastern side of the anticline. Note the numerous intrusions of gabbro (purple color), including the higher regions of Marys Peak. The gray box shows the area of the detailed map in the next chapter.

Stratigraphic Section in Marys Peak Region

Period	Epoch	Age (Million Years)	Formation	Description
Qua-ternary	Holo-cene	< 0.012	**Alluvium**	Layers of mud, silt, sand and gravel deposited in modern lakes, streams and floodplains.
Tertiary	Oligo-cene	35 – 30	**Gabbro Intrusions** (Stops 4, 5 & 6)	Coarse-grained igneous rock intruded into older rocks as dikes cutting across layers and sills parallel to layers.
Tertiary	Eocene	50 – 35	**Spencer Formation**	Sandstone, siltstone and shale layers deposited on the ocean floor above the basalt.
Tertiary	Eocene	50 – 35	**Yamhill Formation**	Sandstone, siltstone and shale layers deposited on the ocean floor above the basalt.
Tertiary	Eocene	50 – 35	**Tyee Formation** (Stops 1 & 3)	
Tertiary	Eocene	55 – 50	**Siletz River Volcanics** (Stop 2)	Basalt erupted in the ocean as massive layers and pillow structures.

The rock layers on and near Marys Peak are young by geological standards, having formed during the past 55 million years. Layers encountered at the six road-log stops are highlighted with white and red text.

intrusion keeps the upper regions of Marys Peak high above other peaks in the Oregon Coast Range.

The term *stratigraphic section* refers to the sequence of rock units exposed at the surface in a specific geographic area. The stratigraphic section used here is simplified. Some maps and other publications refer to a "Kings Valley Siltstone" formation between the Siletz River Volcanics and the Tyee Formation. Others talk about the "Kings Valley Member" as a layer that is sometimes present within the upper part of the Siletz River Volcanics. This Kings Valley unit is exposed along the Harlan Road and in the lowermost part of the layers at Stop 3.

Rocks on Marys Peak

The layers seen on a drive up Marys Peak include sedimentary and igneous rocks formed during the past 55 million years of Earth's history. Most of the rocks are of four basic types: a black volcanic rock (***basalt***) that formed as magma poured out on the ocean floor; sedimentary layers of ***sandstone*** and ***shale*** deposited over the basalt; and hard, gray-colored ***gabbro*** that solidified from magma that intruded the sedimentary and volcanic layers. The sedimentary layers include ***siltstone*** that has finer grains than sandstone but coarser than shale. For simplicity, rocks with silt-to-sand sized particles are referred to as "sandstone" in this guide.

 Igneous rocks solidify from magma, which is molten rock that may contain gases and suspended solid material. The terms "magma" and "lava" are

Marys Peak Rocks

Most of the rocks on Marys Peak include two types of sedimentary rocks (coarse-grained sandstone and fine-grained shale) and two types of igneous rocks (fine-grained basalt and coarse-grained gabbro).

sometimes used interchangeably, but differ in important ways. All melted Earth material is *magma*, whereas *lava* is magma that poured out on Earth's surface, forming extrusive ("volcanic") rocks. When magma solidifies below Earth's surface, intrusive ("plutonic") rocks result.

Igneous rocks are classified according to two parameters: texture and chemistry. Texture refers to the size of the mineral grains, which is a function of how quickly the magma cooled. *Intrusive rocks* are coarse-grained because they cooled slowly within the Earth, where mineral crystals had time to develop. Magma that encounters air or water at Earth's surface cools quickly, so that *extrusive rocks* have fine-grained minerals that you generally cannot see with your naked eye.

The chemistry of igneous rocks is commonly related to the amount of silica contained in the rock minerals. Silica is the same material that makes window glass. The mineral quartz is pure silica. Its chemical formula is SiO_2, meaning it has one atom of the element silicon for every two atoms of oxygen. Rocks with high silica content tend to be light-weight and light in color. A familiar example is *granite*, the pink-to-white-colored intrusive rock used as building stone and for kitchen counter tops.

As silica content decreases, minerals generally have higher percentages of heavier elements like iron, forming rocks that are darker and more dense, like those found on Marys Peak. They include fine-grained *basalt* (Siletz

Igneous Rocks		CHEMICAL COMPOSITION			
		Approximate % Silica (SiO_2)			
		70%	60%	50%	40%
TEXTURE	Fine-Grained Extrusive ("Volcanic")	Rhyolite	Andesite	Basalt	
	Coarse-Grained Intrusive ("Plutonic")	Granite	Diorite	Gabbro	Peridotite

Igneous rocks form when molten Earth material (magma) cools. Their classification depends on the size of the mineral grains (texture) and the types of minerals found within the rock (chemical composition). Igneous rocks on Marys Peak include basalt formed on the ocean floor 55 million years ago and gabbro intruded into the basalt and overlying sedimentary layers 35 million years ago.

River Volcanics), formed about 55 million years ago some distance out in the Pacific Ocean. Spectacular globular forms, known as pillows and seen in the middle part of the drive up the peak, demonstrate that much of the basalt erupted and cooled underwater on the ocean floor. The higher part of the peak is a massive intrusion of coarse-grained *gabbro*, formed about 35 million years ago as magma squirted from below and solidified underground within the older basalt and sedimentary layers. Several other peaks in the area, including McCulloch Peak in McDonald State Forest west of Corvallis, are also capped by gabbro. This rock is so hard and resistant to erosion that it is used locally as building stone, including in many buildings on the Oregon State University campus and the Benton County Courthouse. The rock industry informally calls the rock "Willamette Valley Granite" because, like granite, the gabbro contains coarse mineral crystals visible to the naked eye.

Sedimentary rocks form through the cycle of: 1) erosion, chemical or biological activity that forms particles (sediment); 2) transportation of the sediment by water, wind, or ice; 3) deposition of the sediment in oceans, lakes, streams or on dry land; and 4) compaction and cementation of the sedimentary particles as they are buried beneath more sediment.

Eroded particles of rock are known as clasts. Clastic sedimentary rocks (for example, sandstone and shale) are composed of such particles, while non-clastic sedimentary rocks (such as limestone and rock salt) result from particles that precipitate out of solutions during biological or chemical activity.

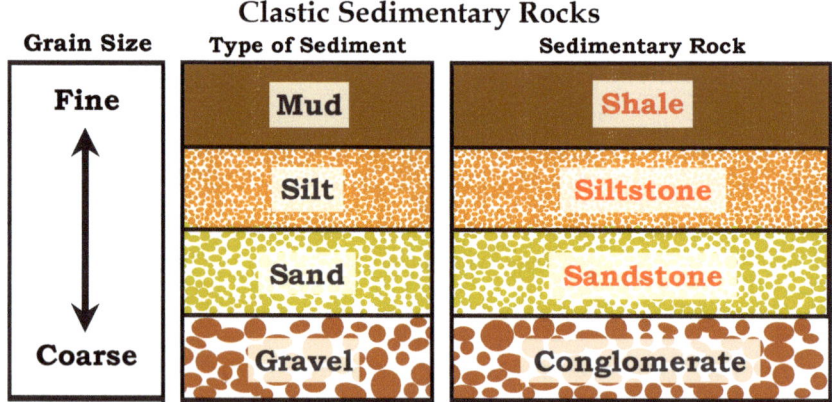

Clastic Sedimentary Rocks

Grain Size	Type of Sediment	Sedimentary Rock
Fine	Mud	Shale
	Silt	Siltstone
	Sand	Sandstone
Coarse	Gravel	Conglomerate

The chart shows rocks formed from the deposition of sedimentary particles (clasts) eroded from other rocks. Marys Peak has prominent sandstone, siltstone and shale layers resulting from the deposition of sand, silt and mud on the ocean floor.

Sedimentary rocks are best understood by imagining their environments of deposition. Gravel is found on the beds of fast-moving streams or along beaches with high wave energy. When buried and cemented, the consolidated gravel layer forms the sedimentary rock *conglomerate*. *Sandstone* is made of sand-size grains, mostly quartz, that were deposited in stream beds or on beaches. In lakes or deeper parts of the ocean, where water is quieter, finer particles of silt and mud accumulate. Burial, compaction, and cementation turn them into *siltstone* and *shale*. In warm climates shell fragments of marine organisms dissolve in seawater, later precipitating out as calcium carbonate. This fine lime mud eventually turns into the sedimentary rock *limestone*. In dry climates where water circulation in an embayment or lake is restricted, high evaporation can leave deposits of *rock salt*.

The sedimentary rocks on Marys Peak are mostly clastic, derived from fine mud deposited on the deep ocean floor (shale) as well as silt and coarser grains of sand from erosion of older rocks on the nearby North American continent (siltstone and sandstone). They were originally deposited as horizontal layers well below sea level. But tectonic forces between the converging plates have lifted the layers to well above sea level and, in places, have contorted and tilted them so that they now are tilted at steep angles.

Geologic Structure

The rock layers on Marys Peak were lifted out of the sea and deformed as they were caught in the vise between the converging North American and Juan de Fuca Plates. In some places they were tilted so much that they now have nearly vertical orientations. The basalt layers, known as the Siletz River Volcanics, erupted off the coast of the Pacific Northwest about 55 million years ago. Similar-age basalt outpourings are found along a line of uplifted terranes in the Coast Range and Olympic Mountains, including the Roseburg Volcanics and Tillamook Volcanics in Oregon and the Crescent Terrane in Washington. Sandstone and shale layers were deposited on top of the basalt during the next 20 million years. The sequence of basalt and overlying sedimentary layers was intruded by low-silica magma about 35 million years ago, forming gabbro dikes (cutting across layers) and sills (parallel to layers).

As the rock layers were caught in the vise between the Juan de Fuca and North American plates, they were shoved upward and eastward along the Corvallis Fault. Movement along the fault line deformed the layers into an upfold known as an *anticline*. The anticline is asymmetric, with sedimentary layers tilted to nearly vertical orientations on the east side and more-gently tilted layers on the west. Stop 1 of the road guide reveals very steep layers,

Geological Development of Marys Peak

West   **East**

55 to 35 Million Years Ago

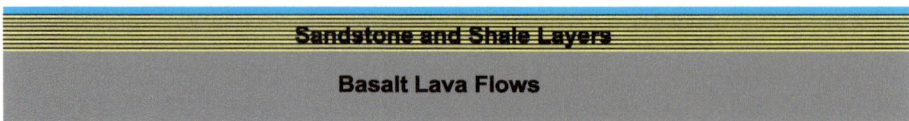

Basalt lava poured out on the ocean floor and was covered by layers of sand and mud that were compressed into sandstone and shale.

35 to 30 Million Years Ago

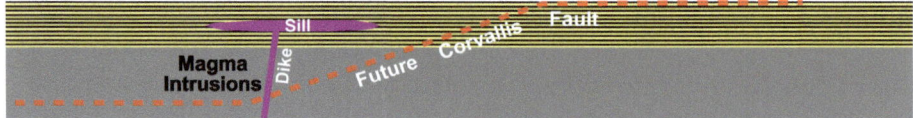

Magma intruded and hardened to gabbro, forming vertical dikes and horizontal sills.

30 to 20 Million Years Ago

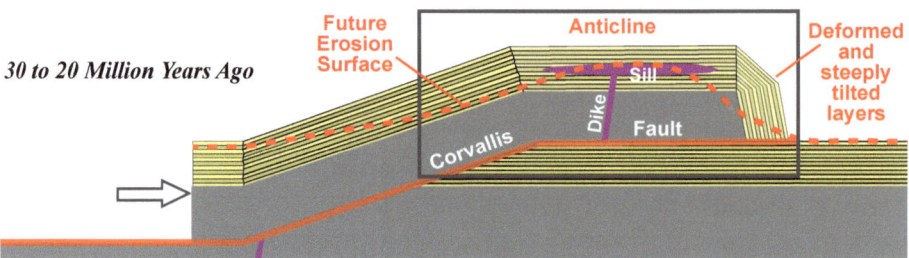

Basalt and sedimentary layers pushed upward and eastward along the Corvallis Fault. Layers deformed and steeply tilted on the east flank of the anticline.

Present

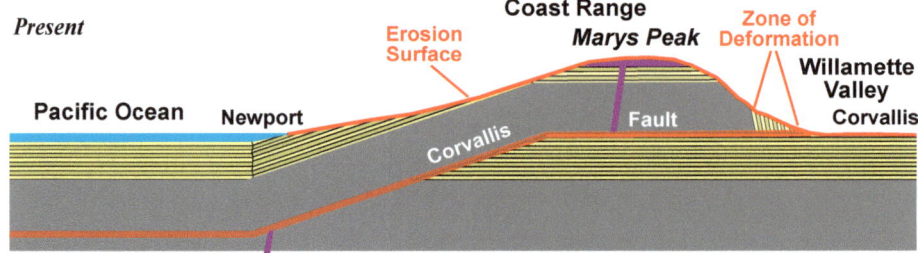

Erosion has removed up to two miles of material above the current erosion surface. Marys Peak remains high because the gabbro sill is hard and has resisted erosion.

(Simplified cross-section modified from Chris Goldfinger, "Evolution of the Corvallis Fault and implications for the Oregon Coast Range," M.S. Thesis, Oregon State University, 1990).

Marys Peak: Highest Point in the Oregon Coast Range

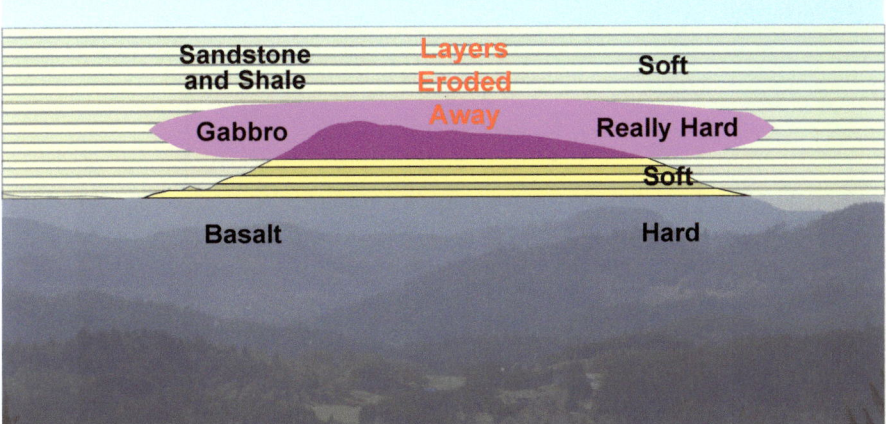

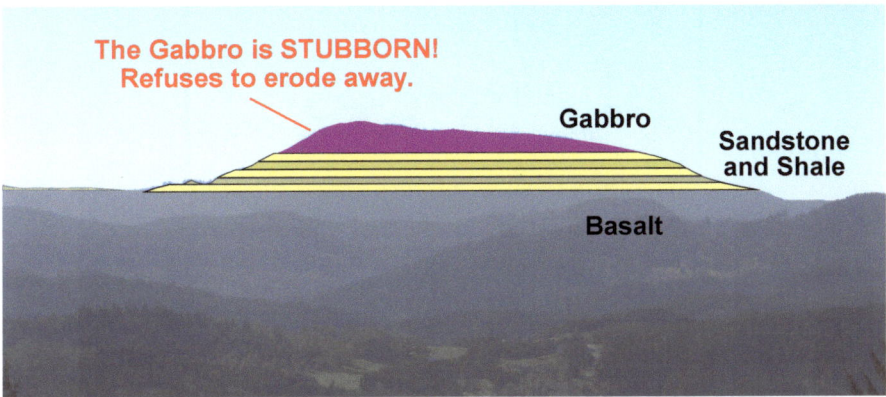

Marys Peak remains so high because the sill of very hard igneous intrusive rock (gabbro) erodes very slowly compared to the other rock layers.

while layers with slight westward tilts can be seen in roadcuts between Marys Peak and the coast. Near the top of the structure, layers are nearly horizontal, as revealed at Stop 3. In addition to being on the east flank of the anticline, Stop 1 lies within a narrow zone of intense deformation where the Corvallis Fault extends to the surface. Wells Creek and Greasy Creek follow this easily eroded zone extending from Alsea Summit, along Highway 34 all the way to the Y-Intersection in Philomath.

As the igneous and sedimentary layers were uplifted, they were simultaneously eroded by the actions of wind, water and ice. The sedimentary rock layers are softer and erode far more easily than the harder and more resistant igneous rock layers. The large block of basalt (Siletz River Volcanics) thus remains as a high area of the central Oregon Coast Range, including the McDonald-Dunn State Forest area west of Corvallis. Locally, Marys Peak remains extremely high—and thus an "Island in the Sky"—because the gabbro intrusion is so hard ("stubborn") that it is taking a long time for it to be dwindled away by erosion.

The Corvallis Fault is the surface along which the Siletz River Volcanics were shoved upward and eastward over the edge of the Willamette Valley, forming a large anticline. The eastern side of the anticline is a zone of deformation within marine sedimentary rocks. The dashed red line shows the approximate eastward extent of the deformed zone (shown in yellow). Nearly vertical layers of sandstone and shale at Stop 1 are within this zone, while Stop 3 reveals the same layers in their original horizontal orientation.

Chapter 3

Geology Road Guide to Marys Peak
A Drive Up to the Bottom of the Ocean

Imagine you could travel to the bottom of the ocean. A drive up Marys Peak is like that imaginary journey. You can see black rock (basalt) that formed as magma poured out on the ocean floor, as well as sedimentary layers (sandstone and shale) deposited over the basalt. Still farther up the mountain, starting at the prominent waterfall at Parker Creek, you first encounter a gray-colored rock (gabbro). This rock has coarse crystal grains that you can see with your naked eye—crystals that grew as magma cooled slowly within the Earth. The top portion of Marys Peak is made of this very hard rock. The gabbro is so resistant to the erosive forces of water and wind that Marys Peak remains nearly 500 feet higher than any other mountain in the region!

The geology road log (page 9) is keyed to distances along Marys Peak Road, where it begins near Alsea Summit on Oregon Highway 34. This intersection is 8.9 miles southwest of the Y-Intersection of U.S. 20 and Oregon 34 in Philomath, and 7.9 miles east of Alsea. The locations of Stops 2, 3, 4 and 5 are keyed to the brown mileage markers along Marys Peak Road, while Stops 1 and 6 on Highway 34 are referenced according to their mileages from the Y-Intersection.

Each stop is also referenced to the simplified cross section of Marys Peak (pages 11 and 34). You can thus see how the types and orientations of rock layers at each stop relate to the broad anticline and the Corvallis Fault. Note that Stop 1 is on the east flank of the anticline and is within the broad zone of deformation caused by the Corvallis Fault (page 36). The other five

Detailed Geology and Road Guide Stops

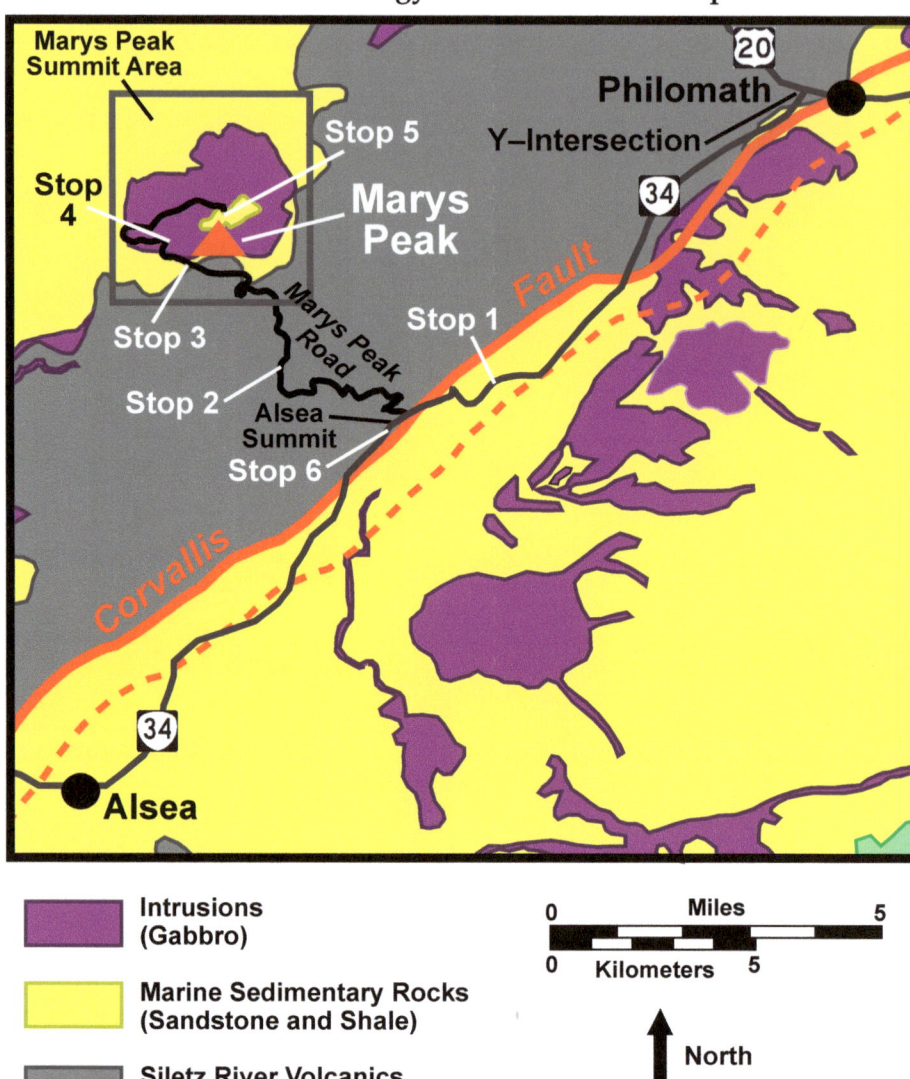

Intrusions (Gabbro)

Marine Sedimentary Rocks (Sandstone and Shale)

Siletz River Volcanics (Basalt)

stops are within a more coherent block of Siletz River Volcanics and its cover of sedimentary and plutonic rocks. This block was uplifted and shoved eastward along the Corvallis Fault without much internal deformation.

Starting at the "Y-Intersection" in Philomath, Highway 34 follows Greasy Creek and Wells Creek toward Alsea Summit. These creeks flow through the broad zone of deformation caused by the Corvallis Fault. The solid red line on the maps and cross section labeled "Corvallis Fault" in not

Marys Peak Summit Area

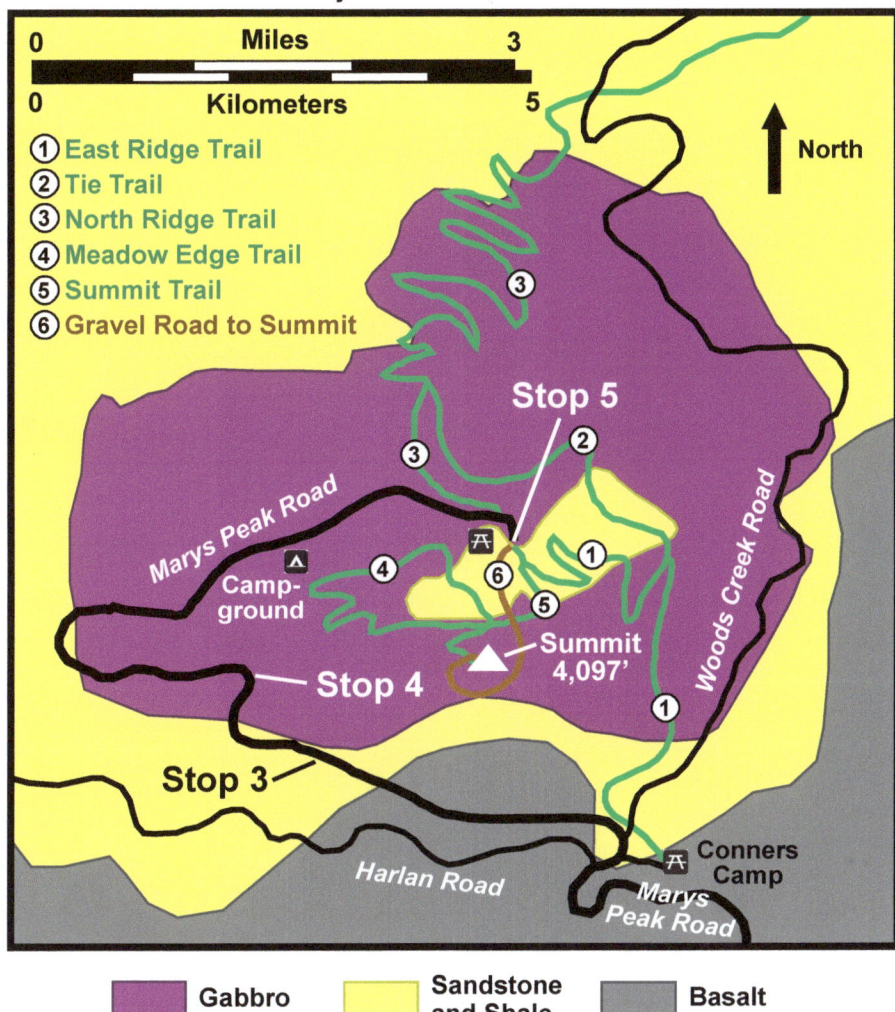

| Gabbro | Sandstone and Shale | Basalt |

The Marys Peak summit area consists of a gabbro sill that intrudes flat-lying sandstone and shale layers. Hiking trails mostly traverse the sill but extend across sandstone and shale at the beginning of the East Ridge and North Ridge trails, and in the area surrounding the parking lot.

intended to represent an abrupt fault line cutting the surface. Rather the line roughly follows the contact between the block of Siletz River Volcanics on the northwest and younger marine sedimentary rocks on the southeast. The dashed red line on the maps approximates the southeast extent of the zone of deformed and steeply tilted rocks on the east flank of the broad anticline.

<u>Stop 1</u>: Steeply Tilted Sedimentary Layers

Oregon Highway 34, 7.4 miles southwest of the Y-Intersection in Philomath and 1.5 miles northeast of Marys Peak Road

GPS Coordinates: 44.4716° N, 123.4841° W

Park in the gravel pullout on left (south) side of road, within the curve of the switchback.

Please use extreme caution parking and pulling back onto highway. For safety, it's recommended that you remain on the south side of the highway and view the roadcut from across the road.

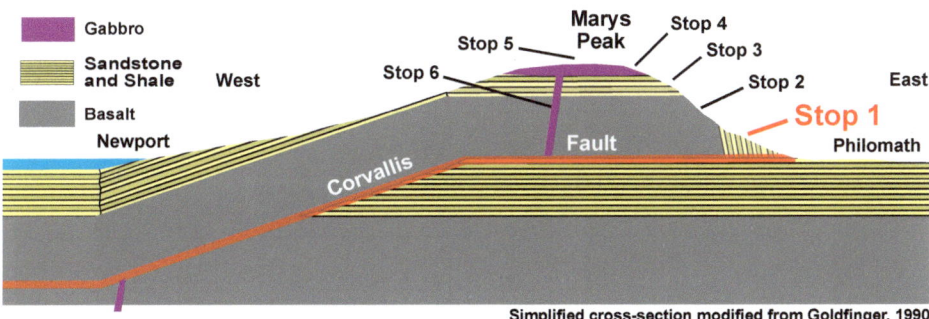

Simplified cross-section modified from Goldfinger, 1990

The initial stop reveals sedimentary layers that were originally deposited horizontally at the bottom of the ocean. The fact that the layers are now steeply tilted and about 1,000 feet above sea level attests to the powerful forces at work where tectonic plates collide.

The layers at Stop 1 are at such a steep angle because they are part of a geologic structure known as an anticline. The anticline formed as the Siletz River Volcanics and their overlying cover of sedimentary rocks and igneous intrusions were caught between the converging Juan de Fuca and North American plates. The squeezing caused the layers to break along the Corvallis Fault and move upward and eastward over the western edge of the Willamette Valley. The anticline is asymmetric, in that layers on the east side are tilted at a much steeper angle than those on the west.

The steep tilting of the layers at Stop 1 could be due to the general tilting on the east flank of the broad anticline, or to more localized folding and

← Steeply tilted layers at the base of Marys Peak reveal Earth's dynamic forces in action.

Tilted Sandstone and Shale

The tilted sedimentary layers at Stop 1 can be viewed most safley from the gravel parking area on the inside of the switchback curve (foreground).

faulting in the zone of intense deformation caused by movement along the Corvallis Fault.

Erosion has removed some of the layers from the top portions of the anticline, so that the lower portions of the Marys Peak Road traverse older basalt of the Siletz River Volcanics exposed in the core of the anticline (Stop 2). But note that the thick gabbro sill (Stops 4 and 5) is so hard and resistant to erosion that it protects a zone of sedimentary layers that remain nearly horizontal close to the top of the peak (Stop 3).

The tilted layers at Stop 1 include several thick layers of sandstone and

Thick Sandstone and Thin Shale Layers

Coarse-Grained Sandstone

Fine-Grained Shale

The thick sandstone layers have coarse grains, primarily of the mineral quartz, and maintain solid blocks. The thin shale layers consist of fine grains of mud and easily break into fragile slabs.

siltstone, separated by thin layers of shale. Such sedimentary sequences are known as *turbidites*. The relatively coarse layers of sand and silt were deposited rather quickly, while the intervening layers of mud accumulated over much longer periods of time. (For safety, it's recommended that you not attempt to cross the busy highway at this stop. Similar, but flat-lying, layers at Stop 3 can be viewed up close and more safely).

Stop 2: Pillow Basalt

Mile 3.7 on Marys Peak Road (12.6 miles from Y-Intersection)

GPS Coordinates: 44.4815° N, 123.5354° W

Park in the large parking lot on the left (west) side of the road.

For the best exposures of pillow basalt, walk ¼ mile west along the gravel road past the gate at the southwest corner of the parking area. Please use extreme caution if visiting the rock quarry because the area is commonly used for target practice by gun owners. For safety, it's recommended that you not approach the rock quarry if you hear gunshots. **You can also see weathered, yet still impressive, pillows in the road cut at the saddle on Marys Peak Road, just beyond the Stop 2 parking lot (accessible a short walk from the same lot).**

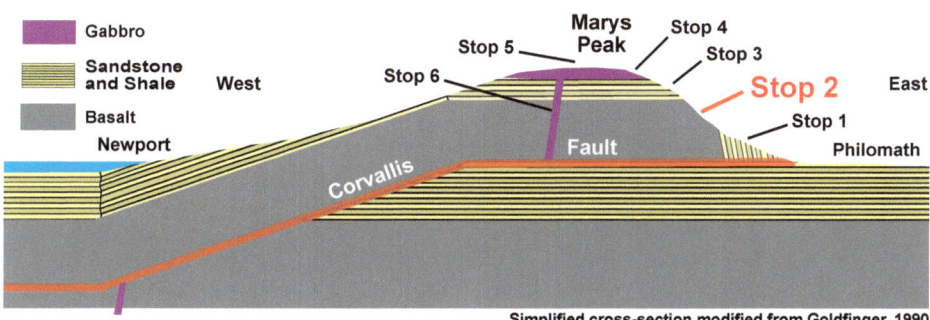

Simplified cross-section modified from Goldfinger, 1990

The spectacular pillow lavas seen on the road up Marys Peak indicate that at least some of the 55 million-year-old Siletz River Volcanics formed on the seafloor and were later lifted out of the ocean. The rock quarry on Marys Peak is one of the best examples of *pillow basalt* found anywhere in the world. It provides details of not only the overall form of the pile of pillows, but also the internal structure developed within pillows as they cooled and hardened.

The pillow form is a very distinctive indication that molten lava cooled and hardened underwater rather than on land. Low-silica (basaltic) magma tends to be very fluid, traveling for long distances over land and forming thin sheets. But when erupting lava encounters water it cools quickly and

← Basalt pillows are evidence that the Siletz River Volcanics formed in the ocean.

Formation of Pillow Basalt

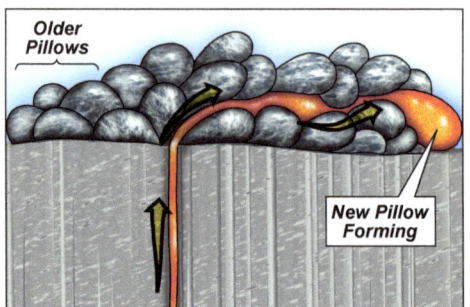

Basalt pillows show that rocks on Marys Peak formed beneath the ocean. As a pulse of hot lava erupts into seawater, it cools quickly and hardens into a pillow that settles on the ocean floor. Other pulses create pillows that pile up on, and fill the gaps between, previously formed pillows. *(From "Earth: Portrait of a Planet," by S. Marshak, ©2001, W. W. Norton and Company, New York).*

assumes a globular form. The globs flatten out under their own weight, thus assuming a "pillow" form. This will not happen on land, in the open air. Thus the presence of pillows is a great indication that lava either flowed from land into water (as it does off oceanic islands such as Hawaii), or that it erupted out of the seafloor (as it does at mid-ocean ridges such as the Mid-Atlantic Ridge and Juan de Fuca Ridge). Pillow basalts almost always indicate that the lava cooled within the ocean, although there are some freshwater exceptions, such as the pillows found adjacent to Wizard Island within Crater Lake in the Cascades of southern Oregon.

The extremely wet climate of the Pacific Northwest causes soils more than 50 feet thick to form. Rock quarries and road cuts temporarily reveal the original appearance of Coast Range rocks, before intense weathering distorts the form, color and chemistry of the rocks. The rock quarry at Stop 2 enables us to see the details of both the exterior and interior of pillows with remarkable clarity. Many of the unbroken pillows have a glassy outer rind. The glassy texture formed because the outer portions of a pillow solidified so

Radial Jointing

The rock quarry on Marys Peak reveals that pillows shrink and crack as they cool and harden. **Left:** The broken pillow shows what the interior of an unbroken pillow might look like. **Right:** Radial fractures (joints) formed within the interior of a pillow as it cooled and shrank.

Zeolite

White crystals of zeolite create a striking contrast with the black basalt at Stop 2.

Pillow Basalt in Roadcut

Although more weathered than the rocks in the quarry, the pillow basalt exposed in the roadcut at the saddle just beyond Stop 2 is another outstanding example of lava that flowed into the ocean.

quickly in the cold ocean water that there was not enough time for crystals to form. The interior remained liquid for a longer time. As the interior cooled and solidified, the pillow shrunk and cracked in a radial pattern. Remarkable examples of *radial jointing* are observed within pillows that were cracked open in the quarry.

The block of basalt on Marys Peak contains white material that often has visible crystalline structure. This material, known as *zeolite*, forms when hot water interacts with the basalt and dissolves some of the silica and other chemicals in the rock. The silica, along with aluminum and varying amounts of other elements such as sodium, potassium and calcium, precipitates out of solution and forms the strikingly white zeolite deposits. Some of these zeolite crystals appear blocky, whereas others reveal long white needles radiating from a central point.

Marys Peak is one of numerous places in the Coast Range where basalt rock is quarried. The reason is all around as you drive up Marys Peak—tall trees! The Pacific Northwest produces much of our nation's timber. Roads are necessary to harvest, replant and manage the supply of trees. Basalt from quarries such as the one at Stop 2 provides much of the surface material for thousands of miles of logging roads in the region.

Stop 3: Flat-Lying Sedimentary Layers

Mile 6.4 on Marys Peak Road (15.3 miles from Y-Intersection)

GPS Coordinates: 44.5003° N, 123.5605° W

Park in the wider shoulder 1/10 mile beyond the layers, on the left (west) side of the road. Walk back along the road to view the flat layering.

Please use extreme caution parking beside and pulling back onto the roadway. For safety, it's recommended that you remain on the west side of the highway and view the roadcut from across the road. If you do cross the road, be sure to look both ways and make sure that everyone in your party gets completely off the road when viewing the outcrop.

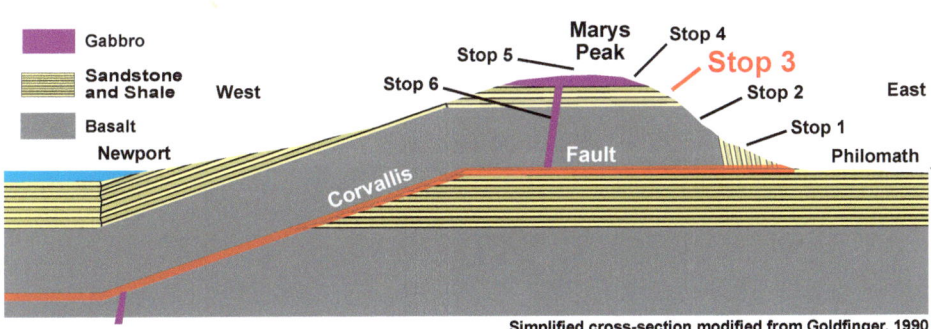

Simplified cross-section modified from Goldfinger, 1990

Most of the sedimentary rocks on Marys Peak result from sand and mud deposited in huge, fan-shaped patterns on the ocean floor. Such deposits of thick sandstone and siltstone layers separated by thin shale layers are called *turbidites* because they were deposited by fast-moving undersea flows of sediment-laden water known as *turbidity flows*. Turbidity flows are often triggered by earthquakes. High volumes of sediment make the flows so dense that they move quickly down the edge of the continental shelf. As a turbidity flow reaches the flat bottom of the deep ocean it begins to slow down, first dropping the coarse, heavy sand, then gradually the finer silt. The very fine particles continue farther out to sea as mud. During the time between turbidity flows, the water on the ocean floor remains calm, so that a thin layer of mud is deposited on top of the sand and silt. A turbidite

← Flat-lying turbidite sequences show periods of alternating fast and slow deposition, represented by thick sandstone and thin shale layers.

Flat-Lying Sandstone and Shale

The flat-lying sedimentary layers at Stop 3 can be viewed most safely from across the road.

Hard Sandstone and Soft Shale Layers

The hard sandstone layers maintain solid blocks. The softer shale layers are fragile and easily split into broken slabs.

Turbidite Sequence:
Thick Sandstone and Thin Shale Layers

Thick Sandstone
Thin Shale
Thick Sandstone
Thin Shale
Thick Sandstone

Turbidite layers can result from severe shaking of the continental shelf during earthquakes. Each thick sandstone layer was deposited on the seafloor in a few days as dense, sediment-saturated water flowed over the continental slope after an earthquake. Each thin shale layer took decades-to-centuries to deposit as mud accumulated in calm ocean waters during the time interval between earthquakes.

sequence commonly has thick layers of sandstone and siltstone (deposited quickly by a turbidity flow) interbedded with thin shale layers (deposited slowly during the long time between turbidity flows).

The layers within a turbidite sequence represent drastically different rates of deposition that may tell us about huge earthquakes in the Pacific Northwest. Sandstone layers are commonly one to five feet thick. They were each deposited in just few days, as a turbidity flow reached the flat ocean floor and slowed down. The thin shale layers are only a few inches thick but each took hundreds of years to deposit, as fine mud settled out of the calm ocean water.

Stop 4: Gabbro Sill

Mile 6.8 on Marys Peak Road (15.7 miles from Y-Intersection)

GPS Coordinates: 44.5041° N, 123.5630° W

Park in the paved area next to Parker Creek Falls.

Please use extreme caution parking and pulling back onto the roadway. For safety, it's recommended that you remain on the east side of the highway. If you do cross the road, be sure to look both ways and make sure that everyone in your party gets completely off the road when viewing the outcrop.

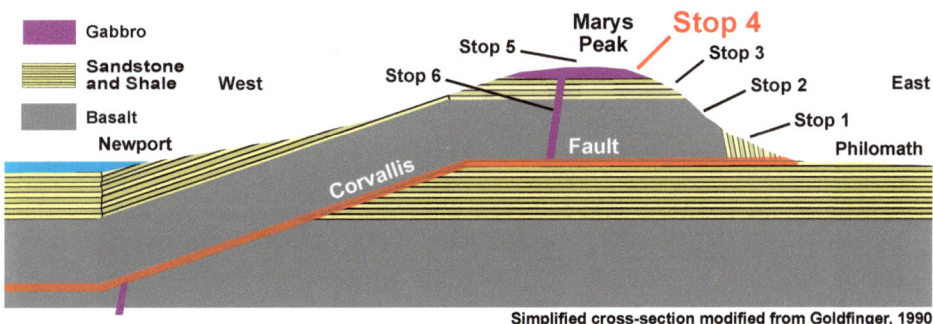

Simplified cross-section modified from Goldfinger, 1990

The rock at this stop has played a key role in the development of Marys Peak as the highest point in the Oregon Coast Range. It's also responsible for the formation and beauty of Parker Creek Falls.

Gabbro is an igneous rock that is similar to basalt. Its dark color is due to the rock's low-silica and high-iron content. The rocks at Stop 2 and Stop 4 thus both have dark, black-to-gray color. They differ in the size of their mineral crystals. **Basalt** formed from magma at or near Earth's surface, where it cooled so quickly that it did not have time for large crystals to form. It is thus a **volcanic rock** and you'd need a microscope to see the crystals. **Gabbro** formed from magma that never made it to the surface. It was insulated deep within the Earth so that large crystals formed as the magma cooled very slowly. It is a **plutonic rock** and the crystals are so large that you can see them with your naked eye or low-powered magnifying glass.

Gabbro is also similar to granite in that its coarse mineral grains are visible to the naked eye. Granite is a rock that is familiar to most people because

← Parker Creek Falls descend over the nearly vertical erosional edge developed on the very hard gabbro sill.

Hard, Erosion-Resistant Gabbro at Parker Creek Falls

The steep walls and presence of the waterfall at Stop 4 reveal that the gabbro exposed in the upper elevations of Marys Peak is strong and highly resistent to erosion. The paved parking area at the falls allows for safe closeup viewing of the gabbro.

Igneous Rocks: Volcanic vs. Plutonic

Fine-Grained Basalt (Stop 2)   **Coarse-Grained Gabbro (Stop 4)**

The rocks at Stops 2 and 4 are both igneous rocks with low silica and high iron composition. But the basalt is volcanic rock that cooled quickly at Earth's surface, whereas the gabbro is plutonic rock that cooled slowly within the Earth.

Yosemite Granite vs. Marys Peak Gabbro

Yosemite National Park Granite

Marys Peak Gabbro

The granite in Yosemite and gabbro on Marys Peak are coarse-grained igneous intrusive rocks. They resist erosion, forming steep rock faces and waterfalls. But granite has more silica and is thus a light pink-to-gray color, while gabbro is higher in iron and is darker, gray-to-black.

it is prominent in iconic places such as California's Yosemite Valley and it is commonly used in building exteriors, monument stones and kitchen counter tops. It is both durable and beautiful. Compared to most other rocks, granite's strength and hardness make it far less susceptible to weathering and erosion than virtually any other Earth material. Granite in Yosemite Valley is valued as extremely hard and reliable climbing rock. It retains nearly vertical faces where spectacular waterfalls descend over 1,000 feet.

<u>Stop 5</u>: Panoramic View of Cascadia Subduction Zone

Mile 9.5 on Marys Peak Road (18.4 miles from Y-Intersection)

GPS Coordinates: 44.5104° N, 123.5450° W

Park in the paved parking area at the end of the road. Restrooms are at the southwest corner of the lot. The subduction zone landscape can be viewed directly from the parking lot. For a broader panorama and views of gabbro outcrops, walk the ½ mile gravel road from the parking area to the top of Marys Peak.

Please stay on the gravel road and designated hiking trails, as the meadows near the top of Marys Peak are a very sensitive and fragile ecosystem.

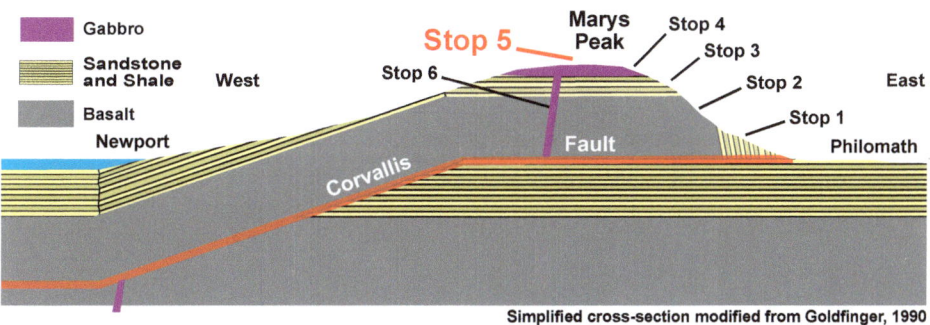

Simplified cross-section modified from Goldfinger, 1990

On a clear day, the top of Marys Peak is one of the best places in the world to see the surface topography of an active subduction zone. Even without walking the extra half-mile to the very top of the mountain, from the parking lot you can see the blue horizon of the Pacific Ocean to the west and the snow-capped Cascade Volcanoes to the east. And on rare days when low clouds fill the Willamette Valley, the high hills of the Coast Range poke through the sea of white, making Marys Peak truly an "island in the sky."

Many elements of the top of Marys Peak reinforce the island in the sky metaphor. The first is the hard bedrock of gabbro. It stands up well against the elements, like the Rock of Gibraltar guarding the entrance to the Mediterranean Sea. The handful of high peaks in the Oregon Coast Range that have very hard bedrock with sub-alpine meadows have been called "balds." In the spring and early summer these open, rocky and often wind-swept

← The gabbro sill comprises the incredibly durable
caprock of Marys Peak.

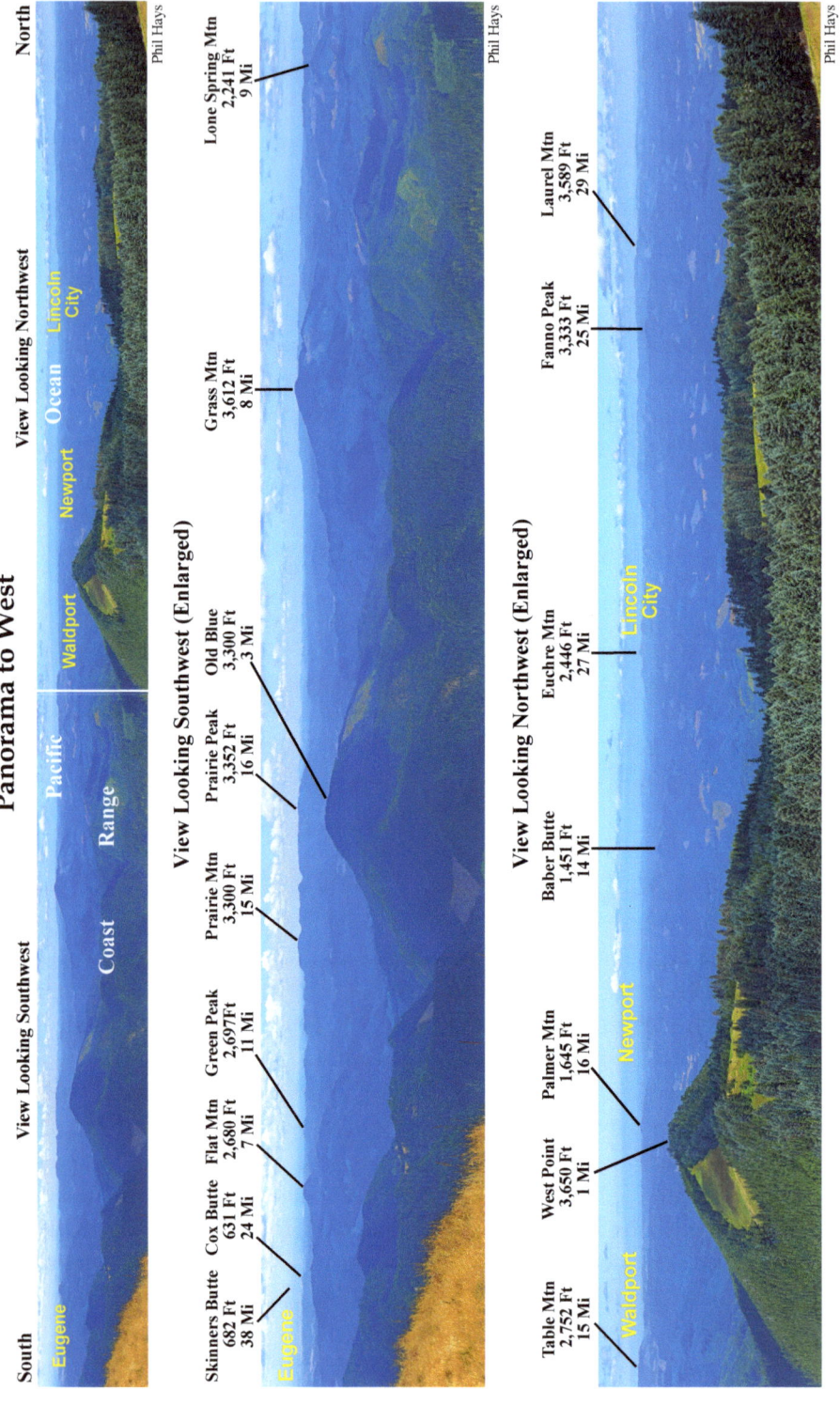

Panorama to West

South

View Looking Southwest

View Looking Northwest

North

Eugene

Coast

Range

Pacific

Waldport

Newport

Ocean

Lincoln
City

Phil Hays

View Looking Southwest (Enlarged)

Skinners Butte
682 Ft
38 Mi

Cox Butte
631 Ft
24 Mi

Flat Mtn
2,680 Ft
7 Mi

Green Peak
2,697 Ft
11 Mi

Prairie Mtn
3,300 Ft
15 Mi

Prairie Peak
3,352 Ft
16 Mi

Old Blue
3,300 Ft
3 Mi

Grass Mtn
3,612 Ft
8 Mi

Lone Spring Mtn
2,241 Ft
9 Mi

Eugene

Phil Hays

View Looking Northwest (Enlarged)

Table Mtn
2,752 Ft
15 Mi

West Point
3,650 Ft
1 Mi

Palmer Mtn
1,645 Ft
16 Mi

Baber Butte
1,451 Ft
14 Mi

Euchre Mtn
2,446 Ft
27 Mi

Fanno Peak
3,333 Ft
25 Mi

Laurel Mtn
3,589 Ft
29 Mi

Waldport

Newport

Lincoln
City

Phil Hays

meadows erupt in an explosion of colorful wildflowers.

The summit area of Marys Peak supports an ecosystem that is the remaining legacy of the last ice age, including the largest noble fir forest left in the western hemisphere. During the last ice age such forests extended to much lower elevations and covered huge portions of the Coast Range. Natural warming since the last ice age has allowed more-temperate Douglas fir forests to creep up to higher elevations. But with accelerated warming due to climate change in recent centuries, it may not be long before the noble fir forests, sub-alpine meadows and stunning wildflower blooms vanish from the Marys Peak island in the sky.

Mt. Hood from the road to the summit. Marys Peak and other Coast Range mountains poke up through low clouds filling the Willamette Valley, like islands in the sky.

Cascadia Subduction Zone: From Marys Peak You Can See it All!

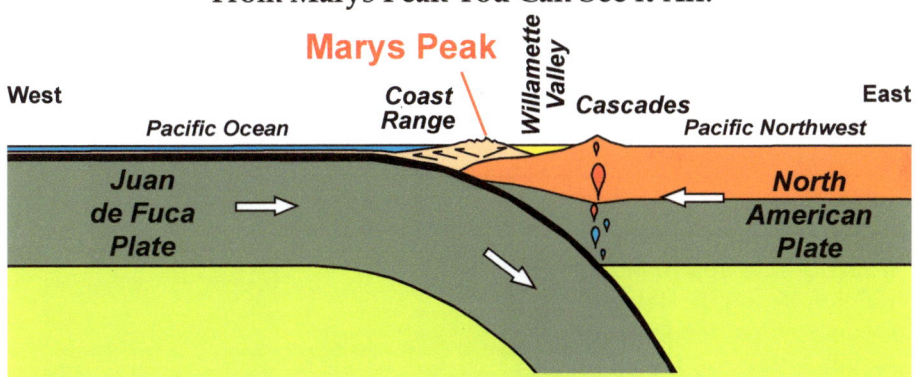

From the upper parking lot (Stop 5) or the very top of Marys Peak you can see the major elements of the Cascadia Subduction Zone. To the west, the Coast Range has rocks scraped off the top of the subducting Juan de Fuca Plate. The Cascade Volcanoes form inland to the east, where the top of the Juan de Fuca Plate reaches about 50 miles depth.

Cascade Volcanoes from Marys Peak

On a clear day you can see most of the major Cascade Volcanoes from the top of Marys Peak.

Mount Rainier

Last Erupted: 1,000 Years Ago

Mount St. Helens

Last Erupted: 2008

Mount Adams

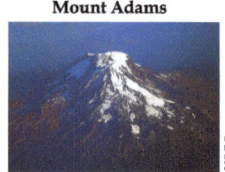

Last Erupted: 3,800 Years Ago

Mount Hood

Last Erupted: 1865

Mount Jefferson

Last Erupted: 6,500 Years Ago

Three Sisters

Last Erupted: 2,000 Years Ago

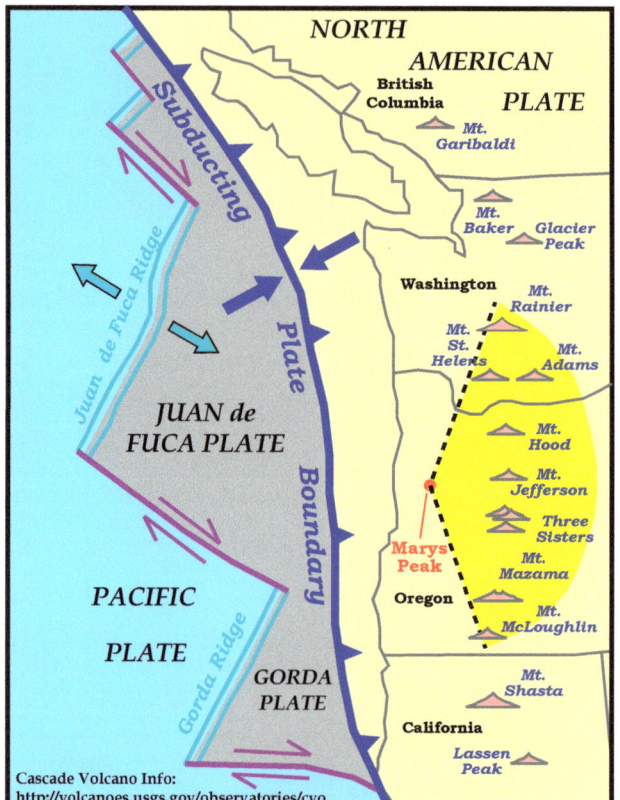

Cascade Volcano Info:
http://volcanoes.usgs.gov/observatories/cvo

The line of sight reveals, from north to south: Mount Rainier, Mount St. Helens, Mount Adams, Mount Hood, Mount Jefferson, Three Sisters, Mount Mazama and Mount McLoughlin.

Mount McLoughlin

Last Erupted: 30,000 Years Ago

Mt. Mazama (Crater Lake)

Last Erupted: 6,600 Years Ago

Panorama to East

North — View Looking Northeast

Cascade Volcanoes — View Looking Southeast

Corvallis
Philomath

Willamette Valley

Eugene

South

Phil Hays

View Looking Northeast (Enlarged)

Mt. Rainier
14,410 Ft
186 Mi

Mt. St. Helens
8,365 Ft
136 Mi

Mt. Adams
12,276 Ft
158 Mi

Mt. Hood
11,239 Ft
110 Mi

Mt. Jefferson
10,497 Ft
89 Mi

Corvallis
Philomath

Phil Hays

View Looking Southeast (Enlarged)

3-Fingered
Jack
7,841 Ft
85 Mi

Mt.
Washington
7,794 Ft
87 Mi

Three Sisters
North: 10,085 Ft
Middle: 10,047 Ft
South: 10,358 Ft
92 Mi

Mt.
Bachelor
9,065 Ft
94 Mi

Diamond
Peak
8,744 Ft
95 Mi

Mt.
Thielsen
9,182 Ft
115 Mi

Mt. Mazama
(Crater Lake)
8,156 Ft
125 Mi

Willamette Valley

Eugene

Phil Hays

A Walk to the Highest Point in the Oregon Coast Range

A stroll up the gravel road to the top of Marys Peak passes through noble fir forests, sub-alpine meadows and wildflowers, all growing on thin soil formed on hard gabbro bedrock.

Marys Peak Meadows and Forests

The meadows and forests on the top of Marys Peak are a legacy of the last ice age. Noble fir forests once extended over larger parts of the Coast Range. Curved tree trunks are due to heavy snow and soil creeping down the steep slopes of Marys Peak.

Marys Peak Bald: Rocks, Meadows and Wildflowers

The hard, resistent caprock of gabbro makes Marys Peak one of the unique hilltops in the Oregon Coast Range known as "balds."

Stop 6: Gabbro Dike

Oregon Highway 34, 9.0 miles southwest of the Y-Intersection in Philomath and 0.1 miles west of Marys Peak Road.

GPS Coordinates: 44.4647° N, 123.5069° W.

Park in the gravel pullout on the left (south) side of the road, near the orange gate just beyond Alsea Summit. Be sure not to block access to the road beyond the gate.

Walk east back toward Marys Peak Road to view the dike in the roadcut on the other side of the road.

Please use extreme caution parking and pulling back onto the highway. For safety, it's recommended that you remain on the south side of the highway and view the roadcut from across the road.

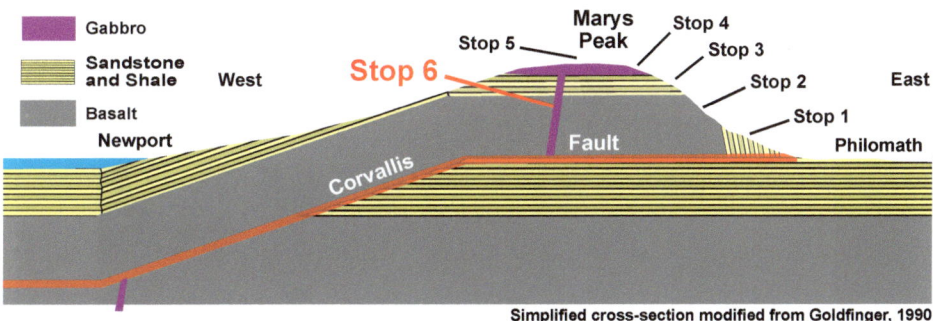

Simplified cross-section modified from Goldfinger, 1990

    The roadcut shows a nearly vertical dike of gabbro that intruded basalt lava flows of the Siletz River Volcanics. Although weathered, the rocks reveal some fundamental geological principles about igneous rock types and orientations, as well as about the relative ages of rocks. First, the gabbro dike is coarse-grained rock, implying that the magma cooled slowly at considerable depth within the Earth, while the basalt is fine-grained, suggesting that it cooled quickly at Earth's surface. Second, the dike cuts through the basalt, indicating that it formed later and is thus younger than the basalt.

    The basalt is part of the Siletz River Volcanics seen in the rock quarry

← Gabbro dikes fill passageways that were the conduits for magma that formed intrusive rock bodies like the one capping Marys Peak.

Younger Gabbro Intruding Older Basalt

The dike at Alsea Summit formed from magma that intruded into much older volcanic rock.

Igneous Rocks in Marys Peak Area

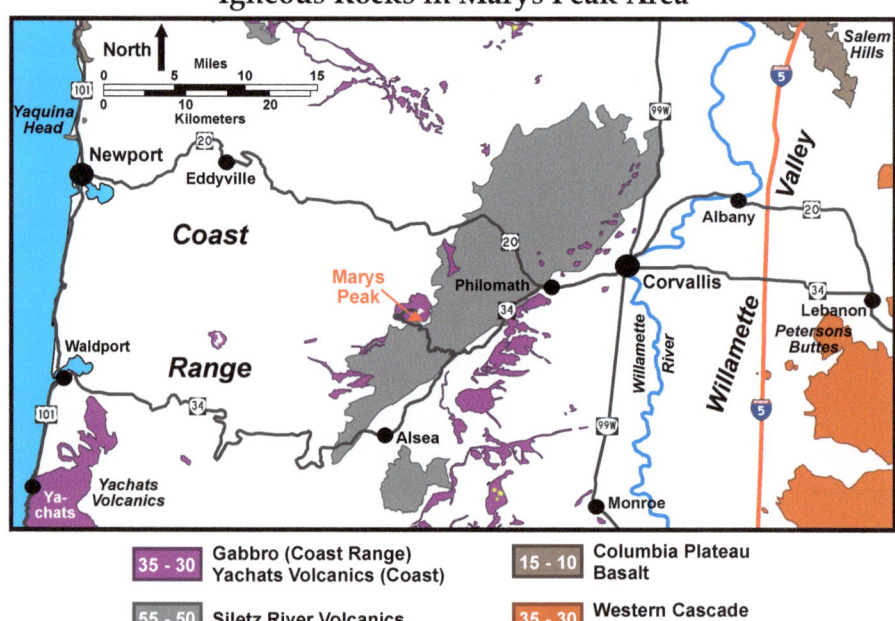

| 35 - 30 | Gabbro (Coast Range) Yachats Volcanics (Coast) | 15 - 10 | Columbia Plateau Basalt |
| 55 - 50 | Siletz River Volcanics | 35 - 30 | Western Cascade Volcanic Rocks |

The white numbers show the approximate time (millions of years ago) when igneous rocks formed in the Marys Peak area. The purple color represents the intrusive gabbro seen at Stops 4, 5 and 6 on Marys Peak, as well as extrusive basalt at Cape Perpetua just south of Yachats.

Intrusive vs. Extrusive Rocks

Coarse-Grained Intrusive (Plutonic) Rock

Fine-Grained Extrusive (Volcanic) Rock

Lava Flow

Volcano

Sill

Dike

Magma

35 Million Year Old Ingeous Rocks

Cape Perpetua

Basalt Lava Flows

Marys Peak

Gabbro Sill

Alsea Summit

Gabbro Dike

at Stop 2. These lava flows formed in the Pacific Ocean about 55 million years ago. The gabbro, also observed at Stops 4 and 5, is much younger, having formed about 35 million years ago. The gabbro thus formed about 20 million years after the basalt. It is thought to be part of the sequence of basaltic magmas that formed at the beginning stages of subduction of the Juan de Fuca Plate about 35 million years ago. Prior to that the subduction occurred farther east, near the region now occupied by the Willamette Valley. As the subduction zone jumped westward to its present position, basaltic magma intruded as gabbro (such as the dike at Stop 6 and the sill at Stops 4 and 5). Some of the magma did surface as basalt lava flows, forming the Yachats Volcanics observed at Cape Perpetua Scenic Area on the Coast, just south of the town of Yachats.

The magma that erupted during the initial stages of subduction of the Juan de Fuca Plate about 35 million years ago formed both intrusive and extrusive igneous rocks in western Oregon. Near the Coast, the Yachats Volcanics formed from magma that reached the surface as basalt lava flows. Magma that cooled before it reached the surface formed dikes such as the one seen at Alsea Summit and sills like the one forming the caprock of Marys Peak.

About the Author

Robert J. (Bob) Lillie is a free-lance writer, illustrator, and interpretive trainer, specializing in communicating park landscapes and their deeper meanings to the public. Bob was a Professor of Geosciences at Oregon State University from 1984 to 2011, where he taught courses in physical geology, oceanography, tectonics, geophysics, geological writing, and public interpretation. He is author of "Parks and Plates: The Geology of Our National Parks, Monuments, and Seashores" (W. W. Norton and Company, 2005) and is a Certified Interpretive Trainer (CIT) through the National Association for Interpretation (NAI). He currently works as a trainer on geology and interpretive methods for the Marys Peak Alliance and the Oregon Master Naturalist Program.

Bob was born and raised in the Cajun Country of Louisiana. He has a B.S. in geology from the University of Louisiana–Lafayette, an M.S. in geophysics from Oregon State University, and a Ph.D. in geophysics from Cornell University. Bob's research focuses on the geological evolution of mountain ranges formed by the collision of continents, including the Himalayas in India and Pakistan and the Carpathians in Central Europe. He is also author of "Beauty from the Beast: Plate Tectonics and the Landscapes of the Pacific Northwest (Wells Creek Publishers, 2015), sold in national parks and forests in the region.

Since 1994 Bob has collaborated with the National Park Service (NPS) on educating the public in geology. He served as a seasonal interpretive ranger at Crater Lake and Yellowstone national parks and John Day Fossil Beds National Monument, and he and his graduate students have written and illustrated geology training manuals for NPS sites across the country. He was presented the 2005 NPS Geological Resources Division award for "outstanding contributions in engaging the National Parks staff and visitors in geoscience." Bob lives with his wife Barb in the Oregon Coast Range near Marys Peak.

← Tiger lilies in the summit meadow of Marys Peak.

www.ingramcontent.com/pod-product-compliance
Lightning Source LLC
Chambersburg PA
CBHW040833180526
45159CB00001B/172

9 781540 611963